Spectrophotometry & spectrofluorimetry

a practical approach

Edited by

C L Bashford

Department of Biochemistry, St George's Hospital Medical School, London, UK

D A Harris

Department of Biochemistry, University of Oxford, Oxford, UK

IRL PRESS

Oxford · Washington DC

IRL Press Limited
P.O. Box 1,
Eynsham,
Oxford OX8 1JJ,
England

British Library Cataloguing in Publication Data

Spectrophotometry and spectrofluorimetry : a practical approach.—(Practical approach series)
1. Biological chemistry—Technique
2. Spectrum analysis
I. Harris,D.A. II. Bashford,C.L.
547.19'2'028 QP519.9.S6

ISBN 0-947946-69-1 (Hardbound)
ISBN 0-947946-46-2 (Softbound)

Preface

The spectrophotometer on the bench is a familiar feature of any biochemistry laboratory. It requires no great skill to use, is sensitive enough to handle materials at physiological concentrations and, best of all, it produces immediate data. As is so often the case, however, familiarity can lead to contempt – too many spectrophotometers are used without sufficient care, are rarely serviced or calibrated and, often, are underused (or even misused) in terms of the facilities they provide.

This book, therefore, is intended to help the reader get the most out of his spectrophotometer and its close relation, the spectrofluorimeter. Applications are described to the characterisation and quantitation of both small and large molecules (photometric assays), to the investigation of intermolecular interactions (ligand binding studies) and to the study of molecular conversions (monitoring chemical reactions) – these both in equilibrium (static) and changing (kinetic) systems.

At every stage, emphasis is placed on the capabilities and limitations of the instrument in use – how to select a machine for a given task, how to check if it is working satisfactorily, and what to do if it fails to produce the expected data. Applications of single, dual and multiple (spectral) wavelength modes of measurement are described, and their respective advantages explained. Chapter 5 also demonstrates the use of the microscope for measurements typically considered the domain of the conventional spectrophotometer or fluorimeter.

It is assumed throughout that commercially available instruments will be used by the reader – those who design a dedicated spectrophotometer for a specific task are unlikely to require this text to help them use it. Some possible workshop modifications of commercial spectrophotometers (e.g. low temperature and rapid mixing attachments) are however described. In some cases commercial instruments are named; the reader should note that in such cases, the specific instrument should be taken as representative of a class and that other manufacturers may well supply comparable instruments.

The editors would like to thank the authors who participated in producing this text, colleagues who read and commented upon it and Mrs B. Bashford for help in producing many of the figures. We are also indebted to Kontron instruments for supplying most of the photographs and diagrams for Chapter 1.

D.A.Harris and C.L.Bashford

Contributors

C.R.Bagshaw
Department of Biochemistry, University of Leicester, University Road, Leicester LE1 7RH, UK

C.L.Bashford
Department of Biochemistry, St George's Hospital Medical School, Cranmer Terrace, London SW17 0RE, UK

J.F.Eccleston
NIMR, The Ridgeway, Mill Hill, London NW7 1AA, UK

D.A.Harris
Department of Biochemistry, University of Oxford, South Parks Road, Oxford OX1 3QU, UK

D.Lloyd
Department of Microbiology, University College of Wales, Newport Road, Cardiff CF2 1TA, UK

R.K.Poole
Department of Microbiology, Kings College, University of London, Campden Hill, London W8 7AH, UK

R.I.Scott
Faculty of Science and Engineering, Polytechnic of Central London, 115 New Cavendish Street, London W1M 8SS, UK

Contents

ABBREVIATIONS **xii**

1. AN INTRODUCTION TO SPECTROPHOTOMETRY AND FLUORESCENCE SPECTROMETRY **1**
C.Lindsay Bashford

General Considerations 1
Absorption of light 1
Emission of light 2
Absorbance Spectrophotometry 2
Types of spectrophotometer 2
Quantitation of absorbance measurements 6
Measurement of absorbance values 7
Common problems of absorbance measurements 8
Specialised applications of absorbance measurements 12
Fluorescence Spectrometry 13
Types of fluorescence spectrometers 13
Quantitation of fluorescence 15
Measurement of fluorescence values 18
Common problems of fluorescence measurements 18
Specialised application of fluorescence measurements 20
Acknowledgements 22
References 22

2. SPECTRA **23**
Robert K.Poole and C.Lindsay Bashford

Introduction 23
Technical Aspects 24
Optical layout 24
Baselines 27
Isosbestic points 28
Wavelength and absorbance calibration 29
Stray light 30
Fluorimeters 30
Choice of Operating Conditions and their Influence on Spectral 'Quality' 31
Wavelength range and light source 31
Effect of spectral bandwidth 31
Scanning speed and instrument response time 33
Temperature 35
Cuvettes 39

A detailed example: recording of a cytochrome difference spectrum (reduced minus oxidised) 41
Factors Affecting the Position of Absorption and Fluorescence Maxima 43
Carbon monoxide 43
Cyanide 44
Photodissociation spectra 44
Derivative Spectra 45
Acknowledgements 48
References 48

3. **SPECTROPHOTOMETRIC ASSAYS** **49**
David A.Harris

Introduction 49
Beer-Lambert law 49
To separate or not to separate 49
Light scattering - clarification of turbid samples 50
General Assay Design 50
Theoretical aspects 50
Practical aspects 52
Type of assay 54
Calibration 55
Spectrophotometric Assay for Proteins 55
General aspects 55
Biuret assay 57
Ultraviolet absorption 58
Lowry method 59
Dye-binding assay 61
Fluorescamine method 63
Assays for other Macromolecules 64
Spectrophotometric assays for nucleic acids 64
Ultraviolet absorption 64
Fluorimetric assay for DNA with Hoechst 33258 64
Fluorimetric assay for DNA plus RNA with ethidium bromide 65
Spectrophotometric Determination of Complex Carbohydrates 66
Assay of Small Biomolecules (Metabolite Assays) 66
Metabolite assays versus assay of macromolecules 66
Design of assay 67
Amount of enzyme required 68
Troubleshooting 69
Determination of Glucose 69
Dye oxidation method 69
NADP-linked assay 70
Choice of method 71
Modifications 71

Rate Methods of Assay 72
'Luminescent' rate assays 73
Determination of ATP using firefly luciferase 74
Assays Including 'Amplification' 77
Metabolite as allosteric effector 77
Metabolite as regenerated co-factor 77
Assay of ($NADP^+$ + NADPH) 78
Reagents 78
Method 79
Modifications 79
Spectrophotometric Measurement of Enzyme Activities 80
Introduction 80
Spectrophotometric measurement 80
Requirements of coupled assay systems 81
Linearity 83
Measurement of ATPase Activity 83
Discontinuous assay 83
Coupled ATPase assay 86
Conclusions 89
References 89

4. MEASUREMENT OF LIGAND BINDING TO PROTEINS 91
Clive R.Bagshaw and David A.Harris

Introduction 91
Applicability of Photometric Methods 91
Advantages and disadvantages 91
Choice of technique 92
Chromophores and fluorophores 93
Practical Considerations 96
Lamp stability 97
Temperature control 97
Photodecomposition 97
Turbidity 97
Measurement of Protein-Liquid Equilibria 98
Principle 98
Selection of wavelength and slitwidth 98
Titration of lactate dehydrogenase with NADH 101
Treatment of Binding Data 107
Rough determination of K_d and n 107
Graphical determination of K_d and n 108
Time-dependent reactions 109
Measurement of actin-myosin interaction during ATP turnover 112
Summary 112
References 113

5. **SPECTROPHOTOMETRY AND FLUORIMETRY OF CELLULAR COMPARTMENTS** **115**
C.Lindsay Bashford

Introduction 115
Apparatus 116
Experimental Design 119
Chromophore selection 119
Characterisation and calibration 119
Artefacts 120
Examples 122
Dependence of mitochondrial oxidation-reduction state on tissue oxygenation 122
Measurement of membrane potential in organelles and in cells 124
Measurement of the pH of cellular compartments 131
Acknowledgements 134
References 134

6. **STOPPED-FLOW SPECTROPHOTOMETRIC TECHNIQUES** **137**
John F.Eccleston

Introduction 137
Rapid reactions 137
Historical development and principle of operation 137
Essential Features of the Instrument 138
Mixing and observation 138
Light source 139
Signal detection 140
The Stopped-Flow Experiment 141
Choice of signal 141
Collection of data 147
Artefacts 149
Testing the Stopped-Flow Instrument 149
Mixing efficiency 150
Determination of dead time 150
Rate Equations 154
First-order reactions 154
Second-order reactions 155
Reversible second-order reactions 157
Two-step binding processes 158
Displacement reactions 159
Competitive binding reactions 159
Analysis of Data 160
Manual methods 160
Computer methods 160
Simulation of reactions 160

Future Prospects 163
Acknowledgements 164
References 164

7. **THE DETERMINATION OF PHOTOCHEMICAL ACTION SPECTRA** **165**
David Lloyd and Robert I.Scott

Introduction 165
Apparatus 165
Current instrumentation 166
Applications and Future Prospects 171
Acknowledgements 171
References 171

INDEX **173**

Abbreviations

ABTS	2,2′-azino-di-(3-ethylbenzthiazoline)-6-sulphonate
ADH	alcohol dehydrogenase
ANS	1,8-anilino naphthalene sulphonate
diO-C_5-(3)	3,3′-dipentyloxacarbocyanine
FA	fatty acid
FCCP	carbonyl cyanide *p*-trifluoromethoxyphenylhydrazone
FP	flavoprotein
GOD	glucose oxidase
G6P	glucose-6-phosphate
GPDH	D-glyceraldehyde-3-phosphate dehydrogenase
G6PdH	glucose-6-phosphate dehydrogenase
HK	hexokinase
Hoechst 33258	[(2-[2-(4-hydroxyphenol)-6-benzimidazolyl-6(1-methyl-4-piperazyl) benzimidazole]
IF-3	initiation factor 3
αKG	α ketoglutarate
LDH	lactate dehydrogenase
MOPS	morpholinopropane sulphonic acid
OA	oxaloacetate
oxonol V	bis[3-phenyl-5-oxoisoxazol-4-yl]pentamethineoxonol
oxonol VI	bis[3-propyl-5-oxoisoxazol-4-yl]pentamethineoxonol
P_i	inorganic phosphate
PBS	phosphate buffered saline
PEP	phosphoenolpyruvate
6PG	6-phosphogluconate
2PGA	2-phosphoglycerate
PGK	phosphoglycerate kinase
PK	pyruvate kinase
PN	pyridine nucleotide
POD	peroxidase
TCA	trichloroacetic acid
thio ITP	6-mercaptopurine riboside-5′-triphosphate
TIM	triose phosphate isomerase

CHAPTER 1

An Introduction to Spectrophotometry and Fluorescence Spectrometry

C.LINDSAY BASHFORD

1. GENERAL CONSIDERATIONS

All biochemicals absorb energy from at least one region of the spectrum of electromagnetic radiation. The energies at which absorption occurs depend on the available electronic, vibrational and rotational energy levels of the molecule. When absorption is from the u.v./visible region of the spectrum (200–700 nm), transitions occur between electronic energy levels, and these electronic transitions form the physical basis for the techniques described in this volume. Spectrophotometry and fluorescence spectrometry (spectrofluorimetry) involve the measurement of these transitions in precise, analytical procedures which permit the characterisation and quantification of (biological) molecules.

A simple appreciation of the fundamental processes occurring when radiation interacts with matter is useful for understanding the operation of spectrophotometers, and is given below. However, a detailed theoretical understanding of these processes is not required for the laboratory application of photometric techniques. Readers interested in these aspects should consult physical chemistry texts.

1.1 Absorption of Light

Molecules absorb energy only when the incident photon has an energy precisely equal to the difference in energy between two allowed states, the photon promoting the *transition* of an electron from the lower to the higher energy state. Before another photon can be absorbed, the excited state must lose this energy and revert to the ground state. Commonly, this reversion is rapid ($<10^{-12}$ sec) and occurs by loss of energy to vibrations and rotations within the same molecule and, by collision, to other molecules (especially the solvent). In short, energy is lost to the environment as heat. The rapidity of reversion is such that, at moderate light intensities, the number of photons absorbed is proportional to light intensity, and constant in time.

If the exciting beam is particularly intense, as it can be with laser light sources, the excitation rate may exceed the rate of decay of the excited state. The number of photons absorbed from a beam of given intensity will thus fall in time as the number of ground state molecules falls – a phenomenon known as *photobleaching*. Such intense sources are thus avoided in the measurements described here; the fraction of molecules in the ground state remains close to one (>99%) and absorption is constant with time.

Another possible cause of photobleaching is a chemical reaction of the excited state.

The chemistry of excited state molecules may differ from that of ground state molecules – they are in general more reactive – and during intense illumination unexpected 'photochemical' reactions may occur leading to incorrect measurements and, at worst, destruction of a valuable sample. However, light sources for absorbance measurements (see below) are rarely sufficiently intense to cause problems of this type. In favourable circumstances the wavelength-dependence of 'photochemical' reactions will provide useful 'action spectra' of complex systems (see Chapter 7).

1.2 **Emission of Light**

In some molecules, particularly rigid conjugated systems, loss of energy from the excited state by vibration or rotation may be slow. In this case, the excited state may lose energy, in addition, by radiative emission i.e. by emitting a photon. If emission is from a singlet excited state, this process is known as fluorescence; if from a triplet state, it is phosphorescence. For observable fluorescence, the lifetime of the excited state must be about 10^{-9} sec, and for phosphorescence it must be about 10^{-3} sec. Clearly, for radiative decay to compete significantly with energy loss as heat, vibrations and rotations within the excited state must be severely restricted to prolong its lifetime.

The competition between radiative and non-radiative decay means that fewer photons are emitted by a collection of molecules than are absorbed; the *quantum yield* (Q_f; Section 3.2) of fluorescence or phosphorescence is less than unity. In addition, during the lifetime of the excited state, some non-radiative loss of energy generally occurs to the environment before emission of the bulk of energy as a photon. This results in the *energy* of the emitted photon being lower than that of the absorbed photon; fluorescence or phosphorescence is at longer wavelength than the corresponding excitation. Factors affecting excitation and emission spectra are discussed more fully in Chapter 2.

While all molecules absorb photons, relatively few fluoresce or phosphoresce significantly at room temperature, so these latter properties are especially useful for resolving minor components in complex mixtures. Furthermore, the high sensitivity of photodetectors and the ability of monochromators or filters to resolve incident from emitted light makes fluorescence, particularly, an exquisitely sensitive analytical procedure. Nanogram amounts of fluorophores can usually be assayed fluorimetrically.

2. ABSORBANCE SPECTROPHOTOMETRY

2.1 **Types of Spectrophotometer**

All spectrophotometers comprise the following elements:

(i) A light source which provides illumination of the appropriate wavelengths. The most common lamps used are tungsten–halogen, for use between 350 and 900 nm, and deuterium, for the u.v. region (200–400 nm). Arc lamps, either of xenon or of mercury, usually contain lines of too great an intensity or fluctuate too much to be commonly employed in absorbance spectrophotometers.

(ii) A device, usually a monochromator or an optical filter, which selects the precise wavelength of interest. In most instruments wavelength selection occurs between the lamp and the sample; in a few instruments, such as the Hewlett-Packard 8450,

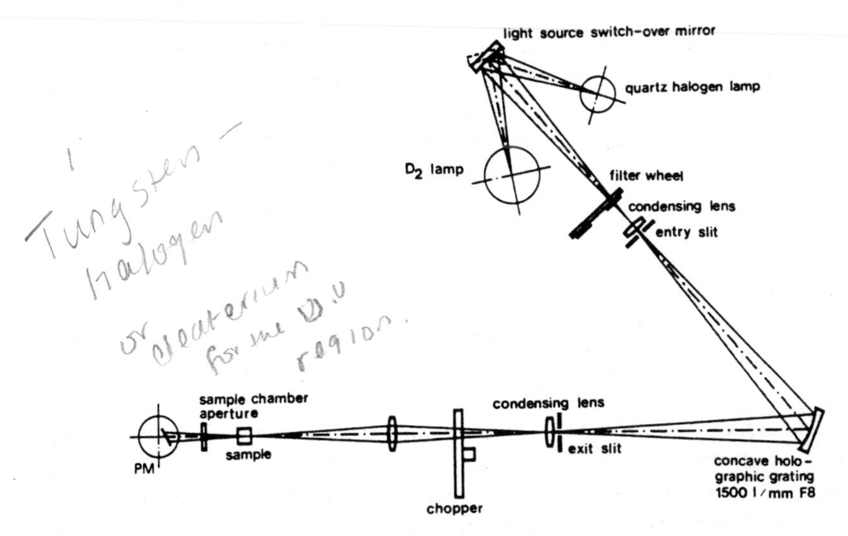

Figure 1. Optical diagram of a single beam spectrophotometer (UVIKON 610/710). PM represents the photomultiplier tube.

wavelength selection occurs between the sample and the detector (an arrangement described as reversed optics).

(iii) A compartment to house the sample to be studied.

(iv) A detector, usually a photomultiplier or a silicon diode, which measures the amount of light transmitted by the sample.

Commercially available photometers incorporate all these features in three main configurations.

2.1.1 *Single Beam Instruments*

These are the simplest type of spectrophotometer. A typical optical diagram of such an apparatus is shown in *Figure 1*. The chopper allows light to illuminate the sample (and the photodetector) intermittently (at a known frequency) and allows the incorporation of a.c. amplifiers into the electronic circuits. Such amplifiers have a better performance than the d.c. devices used if the chopper is omitted. Single beam machines have a single position for sample and reference material. The apparatus is zeroed and standardised with the reference material in the sample position and this is then removed before the sample is studied. Such apparatus is useful for routine assays, for example those described in Chapter 3, where measurements are required, at a single wavelength, of samples and standards. The most important requirement in single beam instruments is that the source output be stable, as changes in transmitted intensity due to variations in source intensity are not compensated.

2.1.2 *Double Beam Instruments*

Corrections for variations in source intensity can be made automatically if the excitation beam is divided between reference and sample materials. This is the strategy adopted by double beam instruments. An optical diagram of such an apparatus is shown in *Figure 2* and a three-dimensional view of the same apparatus is shown in *Figure 3*.

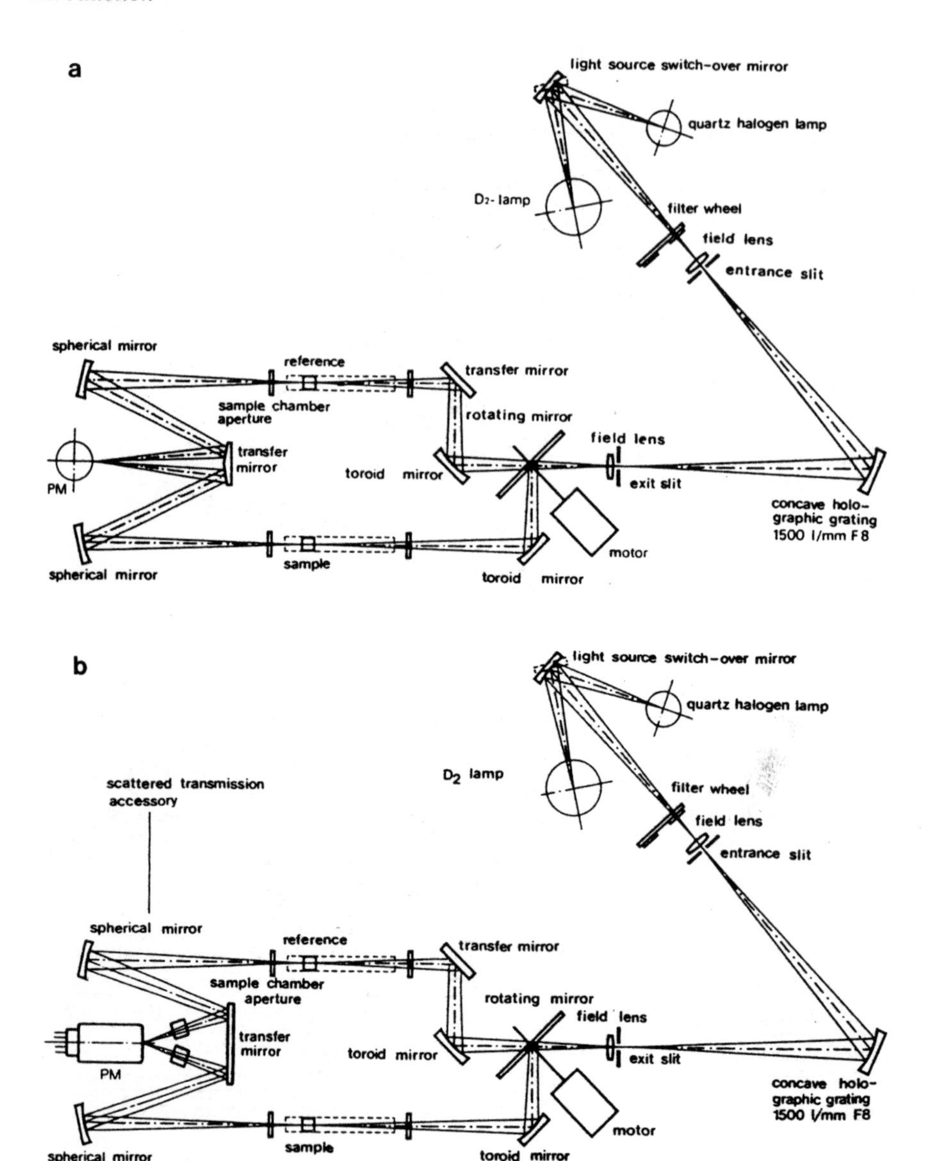

Figure 2. Optical diagrams of double beam spectrophotometers. (**a**) UVIKON 810/860 spectrophotometer with the sample and reference cuvettes in the conventional position. (**b**) UVIKON 810/860 spectrophotometer modified for use with turbid samples. Note that the cuvettes are placed much closer to the photomultiplier (PM) in this configuration.

The essential point is that light of the same wavelength illuminates both the sample and the reference material. In the system illustrated in *Figure 2* the beam is switched from sample to reference by the chopper; the optics for both the sample and the reference chambers are focussed onto the same area of the photodetector (to ensure that each is monitored with the same sensitivity) and a signal from the chopper instructs the electronics as to whether the sample or the reference position is being interrogated. It is

a

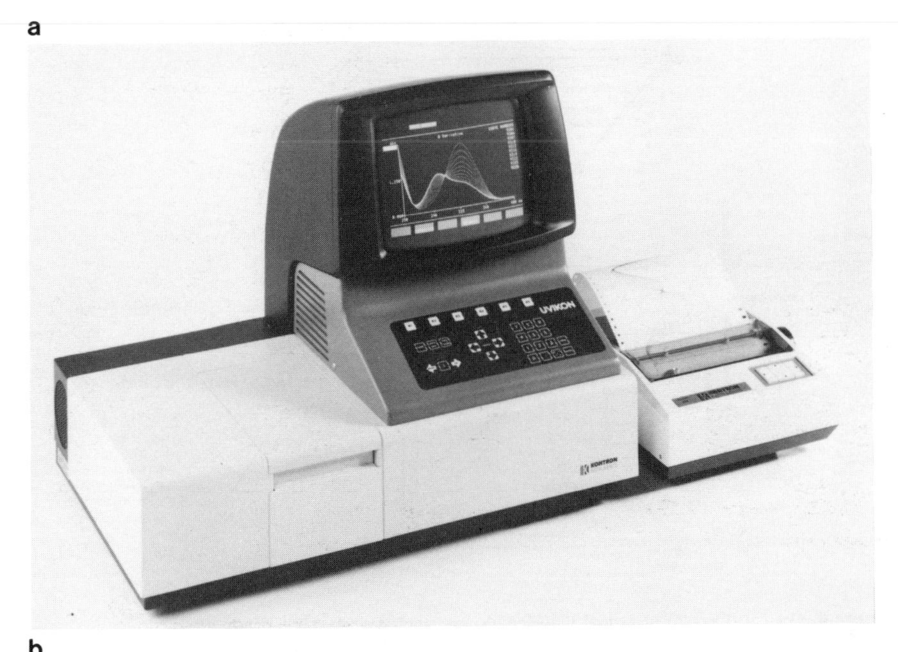

b

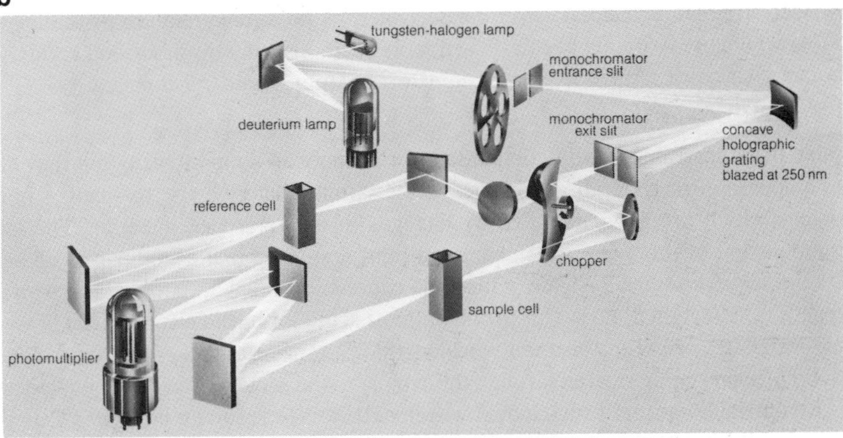

Figure 3. (**a**) A high performance UV/VIS Spectrophotometer using latest technology (UVISOFT-keys and large internal memory to make the instrument very logical to the user. (UVIKON 860). (**b**) Three-dimensional optical diagram of a double beam spectrophotometer (UVIKON 810/860); a modern microprocessor controlled apparatus.

thus possible to compare sample with reference at the frequency (usually 50 Hz) dictated by the chopper.

The use of a chopper limits the kinetic resolution of such devices, changes occurring on a faster time scale than the chopping frequency being unacceptably ‘averaged’. This can be overcome by dividing the beam in space (as opposed to in time) and using a different detector for the reference and sample chambers. Such double beam ‘in space’ devices have very good kinetic resolution (and can be used to follow events in the picosecond time range) but their optical resolution is compromised by the need to provide

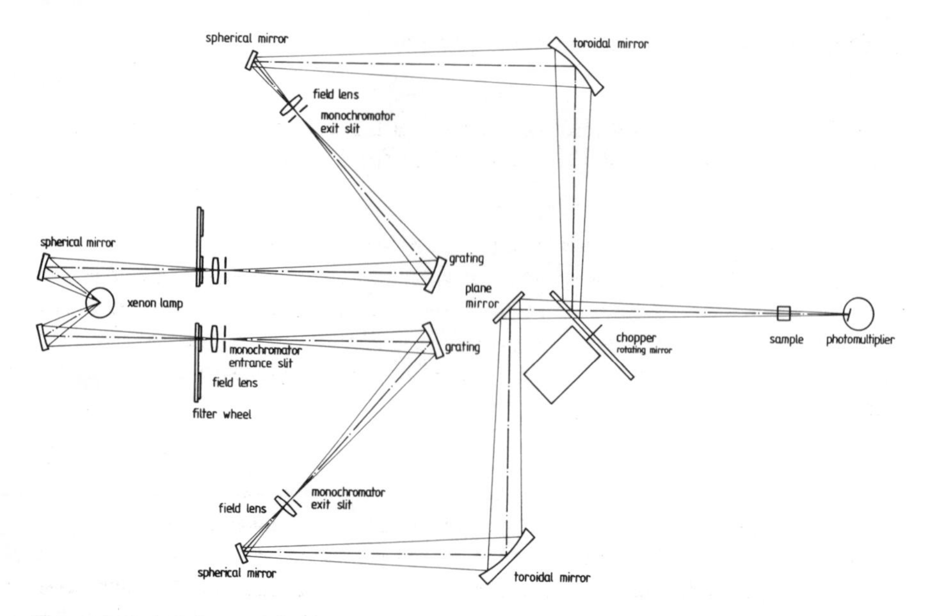

Figure 4. Optical diagram of a dual wavelength spectrophotometer (UVIKON).

very well matched detectors. Unfortunately there is no agreed naming procedure for devices which use two light beams. The double beam instruments described in this section can also be described as 'dual beam' or as 'split beam' machines.

2.1.3 *Dual Wavelength Instruments*

Double beam instruments will compensate for changes in lamp intensity but they cannot compensate for generalised changes in the sample that give rise to changes in absorbance which are not parallelled by changes in the reference material. The most common event of this type is the change in light scattering that accompanies changes in volume of membrane vesicles, which can cause substantial apparent changes in absorbance (see Section 2.4.1). This problem, most acute for turbid samples, can be overcome by using a reference beam of wavelength different from that of the sample beam with both beams being steered through the sample to the same region of the photodetector. An optical diagram of such a dual wavelength device is shown in *Figure 4*. In this apparatus it is important that the light path of each beam is as near identical as possible. As with double beam machines the two wavelengths may illuminate the sample alternately (dual wavelength in time) or may be directed orthogonally to appropriately placed, separate detectors (dual wavelength in space). The latter have a specialised use in high-resolution kinetic studies, but are not commonly found in biochemistry laboratories.

2.2 **Quantitation of Absorbance Measurements**

The absorbance (A) of a sample depends on the ratio of the intensities of the illumination falling on the detector in the absence (I_o) and the presence (I_t) of the sample according to the following relationship:

$$A = \log_{10}(I_o/I_t) \qquad \text{Equation 1}$$

The transmittance (T) of the sample is given by:

$$T = I_t/I_o \qquad \text{Equation 2}$$

Transmittance can have any value between 0 and 1 and is often quoted as a percentage (i.e. varying between 0 and 100). Absorbance and transmittance are related thus:

$$A = -\log_{10} T \qquad \text{Equation 3}$$

The most useful relationship in absorbance spectrophotometry arises from the combination of Lambert's Law, which states that each layer of equal thickness of an absorbing material absorbs an equal fraction of the radiation traversing it, and Beer's Law, which states that the absorbance of a solution is proportional to the concentration of the absorbing solute. The combined Beer–Lambert relationship can be expressed mathematically:

$$A = \epsilon.\ [\text{solute}].\ l \qquad \text{Equation 4}$$

where ϵ is a constant whose value characterises the particular solute and which is known as the molar absorptivity (it is sometimes referred to as the extinction coefficient), and l is the length of the light path through the sample, which is usually 1 cm, but may vary depending on the type of cuvette employed. A, being a logarithm, is dimensionless; l has units of length; ϵ has units of reciprocal concentration and reciprocal length ($\text{l mol}^{-1}\ \text{cm}^{-1}$). The molar absorptivity is the absorbance of a 1 M solution measured in a 1 cm pathlength cell. In some communications the molar absorptivity (ϵ) is quoted in units of $\text{cm}^2\ \text{mol}^{-1}$. The reader should note that a variation in the units used for ϵ leads to a difference of 1000-fold in its numerical value, and take care to use the appropriate value in calculations. Thus:

$$\epsilon = 10\ \text{l mol}^{-1}\ \text{cm}^{-1} = 10000\ \text{cm}^2/\text{mol} \qquad \text{Equation 5}$$

The wavelength dependence of absorbance (absorption spectrum) and molar absorptivity at wavelength λ (ϵ_λ) are parameters that uniquely characterise particular molecules and can be used for both identification and quantification.

2.3 Measurement of Absorbance Values

It is impossible to give a general protocol that covers every type of absorbance spectrophotometer found in biochemistry laboratories. However, there are a number of general procedures that are usually required, and these are discussed in this section. More commonly encountered difficulties and the means by which they may be overcome are also given below. However, it cannot be stressed too strongly that reading the manufacturer's instructions carefully and thoroughly is the only way to extract optimal performance from a spectrophotometer. We strongly recommend that the manuals for spectrophotometers are kept in a secure but accessible location.

2.3.1 *Making an Absorbance Measurement*

After switching the spectrophotometer on and allowing time for the lamp(s) and electronic circuits to warm up and stabilise, the apparatus is ready for use. Modern electronics do not normally require warming up periods, but it is important for the lamps to warm up as the intensity of their output does take time to settle to a steady value. Fifteen minutes is the minimum time for a tungsten–halogen lamp to warm up. Note also that the ventilation of the lamp housing must be clear of obstructions or lamp temperature, and therefore light output, will fluctuate in an erratic and unpredictable manner.

The following sequence of procedures (for a simple, single beam apparatus) is now initiated:

(i) Select the *appropriate wavelength* for the absorption measurement. This may be achieved either by inserting the correct filter, or by turning the monochromator to the correct position, or by instructing the microprocessor which controls the apparatus to select the appropriate wavelength.

(ii) Choose the correct *slitwidth* (if available). A detailed discussion of the effect of slitwidth on absorbance is given in Chapter 2.

(iii) Check the 0% transmittance value. All photodetectors provide some signal to the electronic circuits even when no light falls on the photosensitive surface. This *dark current* must be allowed for in subsequent electronic calculations of absorbance. The 0% transmittance check is the usual means by which this occurs and is a measure of detector output when a shutter prevents light reaching the detector.

(iv) Set *zero* absorbance (or *100%* transmittance) with the *reference* material in position. The reference should match the sample as closely as possible (in terms of medium, temperature etc.) except that the relevant chromophore is omitted.

(v) Replace the reference with the *sample* and read the absorbance (or transmittance) value.

2.4 Common Problems of Absorbance Measurements

2.4.1 *Absorbance not Linear with Concentration*

This condition arises most often at high absorbance values because of a phenomenon called stray light. No matter how sophisticated the apparatus, a small fraction of light in the incident beam will be of wavelengths different from that selected. This can arise from light leaks, which can be blocked by appropriate masking, and because monochromators transmit some specularly reflected light (the grating has the properties of a mirror so this light has the same energy distribution as the source) in addition to the diffracted light. When the selected wavelength is very strongly absorbed, the proportional contribution of the stray light to the transmitted intensity increases because it is of a wavelength that is not absorbed. Thus the intensity of light falling on the detector (I_t) has components due to light of the chosen wavelength (I_c) and the stray light (I_s):

$$I_t = I_c + I_s \qquad \text{Equation 6}$$

and substitution in Equation 1 gives:

$$A = \log_{10}[I_o/(I_c + I_s)] \qquad \text{Equation 7}$$

so that as I_c tends to zero (high absorbance) the measured value of A approaches the constant value of $\log_{10}(I_o/I_s)$. It is necessary to keep $I_c \geq 10\ I_s$ for an accurate assessment of absorbance. From a practical point of view it is useful to determine the stray light characteristics of any photometer by measuring the absorbance of a range of standard solutions. Many simple photometers deviate severely from ideality at absorbance values around 2 (1% transmittance); more elaborate (and expensive) apparatus will give linear readings up to absorbance values of 6.

Non-linearity of absorbance due to stray light is particularly insidious when the reference material has a very high absorbance at the selected wavelength. Then, although the nominal absorbance due to the solute may be small, the background absorbance is so high that only stray light is reaching the detector. This problem arises both from interfering substances (such as absorbing buffers), in which case the remedy is to remove the interfering agent, and in very turbid suspensions. In the latter case the scattering of the collimated, incident beam is so great that little light reaches the detector. High background absorbances due to scattering are ameliorated by bringing the sample and reference materials as close to the photodetector as possible so that a reasonable fraction of the transmitted light reaches the detector (see *Figure 2b*) and/or by selecting cuvettes with a pathlength of less than 1 cm.

If the opaque sample cannot be diluted to give absorbance readings in the linear part of the standard curve, it is occasionally possible to monitor the absorbance at another wavelength where the absorbance is not so great. This is a useful procedure for some standard assays (see Chapter 3).

2.4.2 *Absorbance Reading is not Stable*

This can arise for a number of reasons:

(i) When the atmosphere is humid and the contents of the cuvette are cold, either the reference or the sample cuvette may 'fog'.

(ii) When a cold solution is placed into a warm apparatus, bubbles may be formed due to the decreased solubility of gases at higher temperatures. Bubbles scatter the incident light and tend to increase the apparent absorbance.

(iii) Drops or smearing of liquid on the outside of the cuvette will cause erratic readings.

Problems (i)–(iii) can be overcome by using properly equilibrated samples and by good spectroscopic practice (1,2).

(iv) If there is insufficient liquid in the cuvette, the light beam may pass through the meniscus at the surface. This can be corrected by adding more liquid or by placing a 'spacer' below the cuvette.

(v) At high absorbance levels, the signal/noise ratio is poor. This can be overcome by diluting the sample or by selecting an alternative wavelength. (See also Section 2.4.1 on stray light).

(vi) If the cuvette is too narrow, light may pass around the solution of interest as well as through it. This problem is most commonly encountered with semi-micro cuvettes; these must be adequately masked and/or a narrow slit width used.

2.4.3 *Cannot Set Absorbance Zero*

This arises when the lamp has failed so that no light is reaching the detector. In this instance replacement of the lamp is the only effective remedy. However, there are two other circumstances which give rise to apparent lamp failure:

(i) The holder for the sample and/or reference material is opaque. For measurements using light of wavelengths less than 350 nm it is essential to use quartz or special glass cuvettes. Ordinary glass is completely opaque to u.v. radiation. If you are

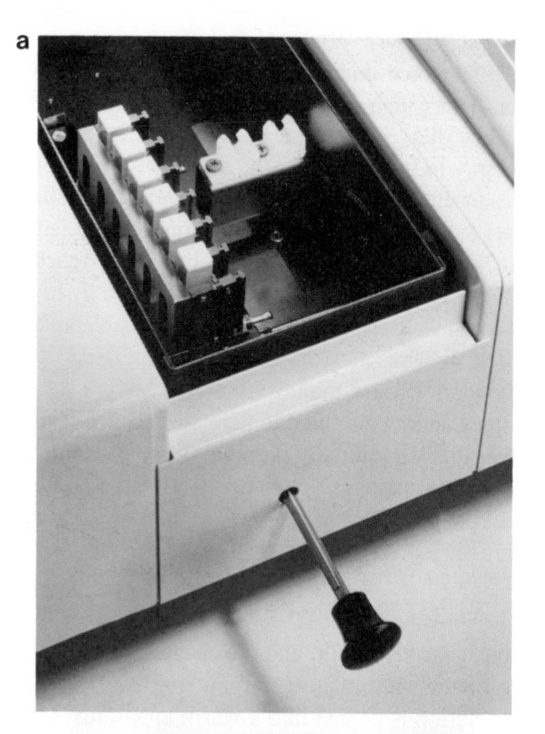

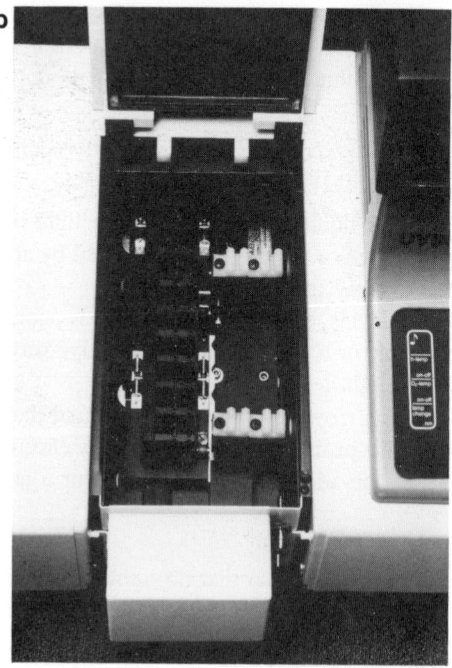

Figure 5. (**a**) Manual cell changer accessory for the rapid measurement of 6 samples in the UVIKON Spectrophotometer. (**b**) View into sample compartment of UVIKON with automatic cell changer to measure 6 samples against 6 references in a UVIKON Spectrophotometer.

a

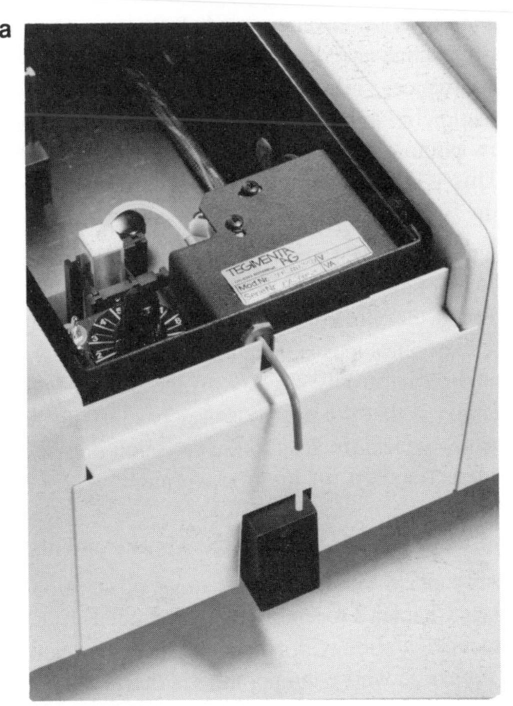

b

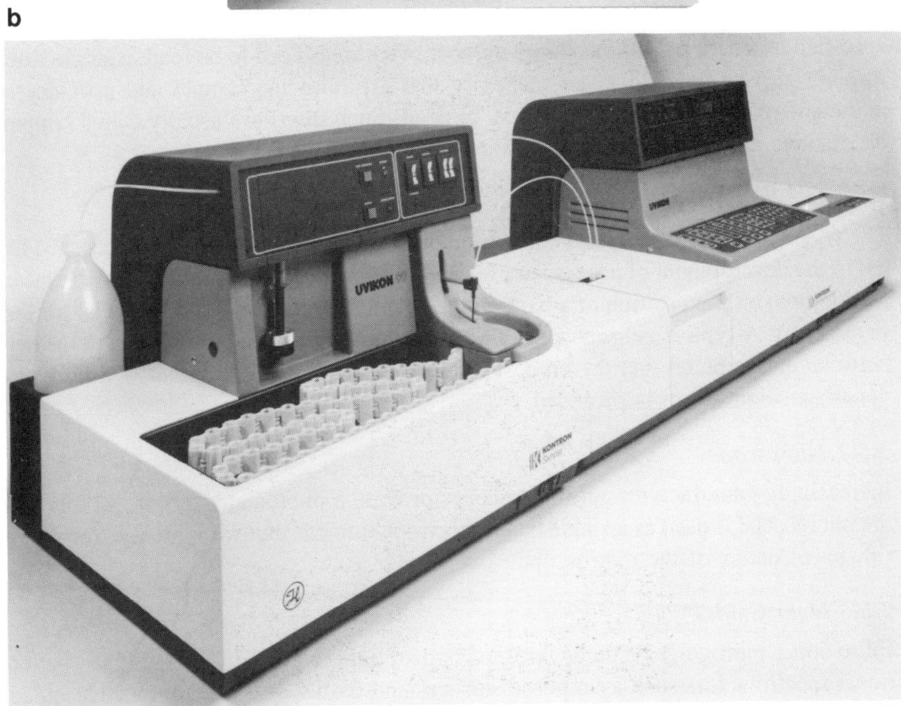

Figure 6. (**a**) Sipper accessory to automatically aspirate the sample into the flow cell. (**b**) Automatic sampling unit allowing the unattended measurement of up to 99 samples.

unsure which material the cuvette is made from, measure its absorbance at 280 nm using air as reference and distilled water as sample; quartz cuvettes will give very low absorbances, glass cuvettes very high absorbances.

(ii) Incorrect selection of lamp type (see Section 2.1). Tungsten–halogen lamps provide very few photons at wavelengths below 350 nm and an alternative source is needed. This is most often a deuterium lamp and it may be labelled 'u.v.' as opposed to 'visible'.

Note that deuterium lamps have some output in the visible region. Indeed the deuterium emission lines at 486 and 656.1 nm can be used as checks for the accuracy of the monochromators. However, deuterium arcs have a low continuum output in the visible region and the sharpness of these lines makes them unsuitable for many absorption experiments as moving the monochromator only slightly off the line leads to an enormous drop in light output. The danger is that if a previous user has left the deuterium source on and the new user selects a wavelength close to the emission line he may not recognise that he is using the incorrect source for his measurement.

2.5 Specialised Applications of Absorbance Measurements

The majority of absorbance measurements are made on solutions or suspensions in conventional cuvettes (see Chapter 2 for more discussion of cuvettes). Typically the cuvettes are held in the correct orientation with respect to the source and detector by a precisely engineered block (*Figure 5*) whose temperature is thermostatically regulated. A motor may enable different positions in the block to be monitored in accordance with user-specified instructions. When a large number of samples need to be read, a single flow-through cuvette coupled to an accessory that aspirates the sample and provides an automatic read-out of absorbance may be used. Such items are usually called 'sipper' accessories and two types are illustrated in *Figure 6*.

2.5.1 *Gel Scanners*

The widespread use of polyacrylamide and other gel electrophoresis techniques (3) has led to the development of accessories for absorbance spectrometers that provide information both on the position of a band in the gel and on the amount of material present in the band. A typical gel scanner is illustrated in *Figure 7*. Essentially the gel is held between quartz plates and driven past the detector such that a read out of absorbance versus position in gel is provided.

2.5.2 *Microscopes*

Increasingly manufacturers of microscopes provide a photometer option. In this case the microscope is used as a single beam spectrophotometer but with the high resolution (in space) optics of the microscope.

2.5.3 *Light Guides*

Fibre optic 'light guides' can be used to conduct light to areas inaccessible to conventional spectrometers such as exposed surfaces and tissues (4). The use of light guides for both reflectance and absorbance measurements enables a study of biological systems *in situ* (see Chapter 5).

Figure 7. Gel scanner accessory for UVIKON spectrophotometers, for the analysis of tube gels or autoradiograms.

3. FLUORESCENCE SPECTROMETRY

3.1 Types of Fluorescence Spectrometers

Fluorescence spectrometers differ from absorbance spectrophotometers in that their design must ensure that none of the incident (excitation) light falls directly onto the photodetector. In practice the number of photons arriving at the photodetector in most absorbance measurements exceeds that in a corresponding fluorescence experiment by a factor of at least 10^6. Furthermore the spectral properties both of the fluorescence excitation (which should be identical with the absorption spectrum; see Chapter 2) and the fluorescence emission are of interest and need to be interrogated separately.

The principal elements of a fluorescence spectrometer are the same as those of the absorption spectrophotometer:

(i) A light source. Fluorescence intensity is proportional to incident intensity (see Section 3.2) so bright sources are commonly employed, usually mercury or xenon arcs. The bright lines of mercury arcs provide excellent excitation but have strong disadvantages for the recording of excitation spectra as the number of photons varies enormously with excitation wavelength. Xenon arcs have much brighter continuum emissions and are widely used in wavelength scanning fluorimeters. However, the presence of bright emission lines still presents problems for the accurate evaluation of absolute or 'corrected' excitation spectra (see Chapter 2). Recently pulsed sources have been introduced in many fluorimeters as these provide intense, short pulses of illumination very suitable for exciting fluorescence. Their main advantage is that their overall power can be much reduced thereby limiting the photochemical production of ozone by the source.

(ii) Filters and/or monochromators to select appropriate excitation and emission wavelengths. Filters are still employed in many fluorescence spectrometers because they allow greater transmission than monochromators and because of the absence (in the presence of appropriate blockers) of higher order effects. It is important in the construction of filter fluorimeters to ensure that the filters are not themselves fluorescent. Thus for example the blocking filters of com-

a

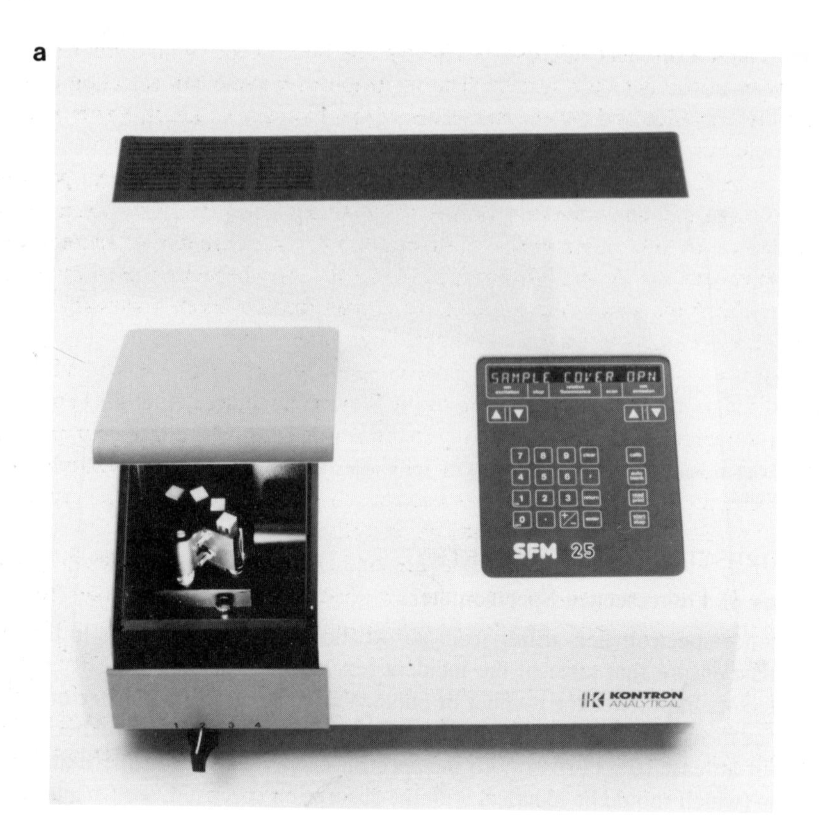

b

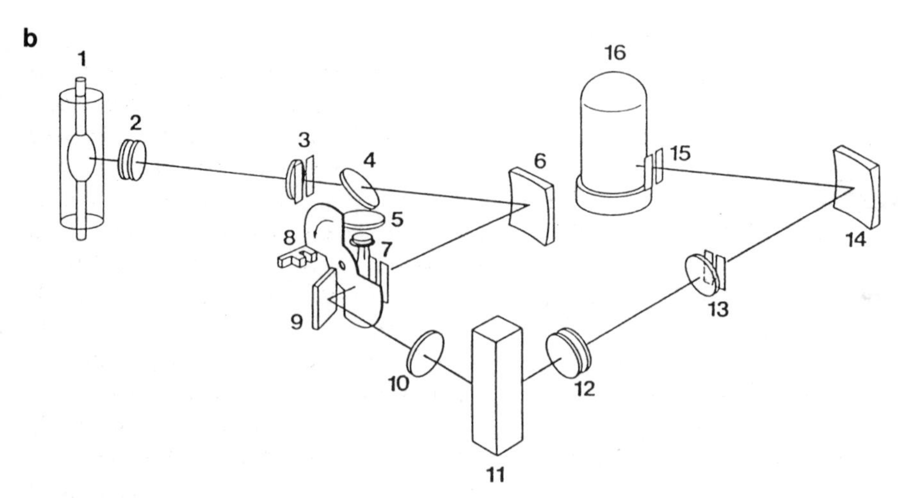

Figure 8. (**a**) Modern scanning spectrofluorometer with 4 position cell changer (KONTRON SFM 25). (**b**) Three-dimensional optical diagram of a modern spectrofluorometer (KONTRON SFM 25). 1: xenon lamp; 2,10,12: collimating lenses; 3,7,13,15: selectable slits at the exits and entrances of the monochromators; 4: beam splitter; 5: reference detector (photodiode); 6,14: diffraction gratings; 8: chopper; 9: plane mirror; 11: sample in cuvette; 16: photomultiplier.

mercial fluorescence microscopes have a low fluorescence which limits the sensitivity of the detector system. The inclusion of wavelength selection devices on the excitation and the emission sides of the fluorescence spectrometer reduces the photon flux arriving at the detector.

(iii) A compartment to house the sample. This must include entry and exit ports for the excitation and emission beams respectively and rigorously exclude environmental light (a major source of stray light in fluorescence measurements).

(iv) A photodetector. As in absorbance machines this may be either a photomultiplier or a silicon diode. Fluorescence microscopes often use diode arrays for photon detection which can have exceedingly high sensitivity.

The majority of commercial fluorescence spectrometers place the photodetector orthogonally to the excitation beam (*Figure 8*). This arrangement minimises contributions from Rayleigh and Raman scattering to the detected signals, although in practice such occlusion is never complete. Even with completely transparent solutions and rigorously cleaned cuvettes it is usual to detect a substantial signal when the excitation and emission wavelengths are set to the same value.

Commercial fluorescence spectrometers are almost always of a single beam type as illustrated in *Figure 8*. Machines such as the one illustrated here provide 'ratio' records by splitting the excitation beam and diverting a small (usually 10%) fraction to a second detector which controls the sensitivity of the primary detector such that if the illumination intensity falls the sensitivity increases and *vice versa*. It should be stressed that such an instrumental 'correction' procedure does not provide properly corrected spectra (see Chapter 2) although the emission spectra recorded in the ratio mode are closer to the true emission spectrum and less machine-dependent.

3.2 **Quantitation of Fluorescence**

Fluorescence spectroscopy lacks a relationship equivalent to the Beer–Lambert Law for absorbance spectroscopy. Consequently fluorescence intensities unfortunately reflect the apparatus in which they were recorded as much as the nature of the sample. Perhaps the least arbitrary parameter of fluorescence that can be measured is the quantum yield (Q_f) which may be defined as the fraction of photoexcited molecules which lose their excess energy as fluorescence. Q_f can take on values between 0 (for non-fluorescent molecules) and 1. A detailed description of the measurement of quantum yields is beyond the scope of the present discussion but useful treatises on the subject are available (1,5). Q_f values depend very strongly on environmental factors such as temperature, solvent composition, local polarity and presence of quenching agents; in this regard they are much more variable than the extinction coefficients useful in absorbance spectrometry.

The theoretical dependence of fluorescence on chromophore concentration can simply be derived from the Beer–Lambert Law (Section 2.2) which can be rewritten as:

$$I_t = I_o . 10^{-\epsilon . s . l} \qquad \text{Equation 8}$$

where I_o and I_t are the intensities of the incident and transmitted light beams, ϵ is the molar absorptivity, s the molar concentration of solute and l the path length. The intensity of absorbed light (I_a) is

$$I_a = I_o - I_t \qquad \text{Equation 9}$$

and combination of Equations 8 and 9 gives:

$$I_a = I_o.(1 - 10^{-\epsilon.s.l}) \qquad \text{Equation 10}$$

The intensity of fluorescence is given by:

$$I_f = I_a.Q_f \qquad \text{Equation 11}$$

If we substitute for I_a in Equation 11 using the value given in Equation 10, expand the exponential term and ignore the higher order terms (a reasonable assumption when $\epsilon.s.l. < 0.05$) we obtain:

$$I_f = 2.303\ .I_o.Q_f.\epsilon.s.l \qquad \text{Equation 12}$$

Thus in dilute solutions fluorescence is proportional to: (i) the incident intensity (bright sources give more fluorescence than dim ones); (ii) the absorptivity; and (iii) the quantum yield. Thus, under these conditions fluorescence intensity is directly proportional to concentration of the fluorophore. However, the assumptions apply only for solutions with absorbance values of less than 0.05. At higher values a non-linear dependence of fluorescence on concentration is expected and appropriate calibration curves must be constructed for analytical measurements.

3.2.1 *Inner Filter Effect*

The dependence of fluorescence intensity on concentration can be non-linear for reasons additional to the mathematical ones discussed above. Of the additional mechanisms the most important, and the most easily overlooked, is the inner filter effect. This can arise either because the excitation light is strongly absorbed by the solution containing the fluorophore or, more rarely, when the emitted light is strongly absorbed.

For those unfamiliar with the inner filter effect the following practical demonstration is instructive.

(i) Make a 1 mM solution of a very fluorescent material such as fluorescein or 9-aminoacridine.
(ii) Place the solution in a cuvette in a fluorimeter (or spectrophotometer).
(iii) Set the excitation (incident) wavelength to the appropriate maximum.
(iv) Open the excitation shutter, close the emission shutter and switch off the room lights.
(v) Open the sample compartment and look down on the cuvette.
(vi) Observe where the fluorescence arises.

Fluorescence is easily recognised as having a longer wavelength than the excitation beam, and for 9-aminoacridine the excitation at 410 nm is dark blue and the emission is light blue. Now dilute the sample two-fold and again observe the fluorescence. As the material is successively diluted you will observe the deeper and deeper penetration of the excitation beam into the cuvette (*Figure 9*). It is surprising how many dilutions are possible before the fluorescence is no longer visible to the naked eye. When the detecting system of the fluorimeter is used to record the fluorescence of such a series of dilutions, the first dilutions lead, paradoxically, to a fluorescence increase because the detector 'looks' at a part of the cuvette not reached by the excitation beam (*Figure 9*). Only in dilute samples is the expected linear relationship between fluorescence and concentration observed (*Figure 9*). As indicated above it is good working practice to arrange for the absorbance of the sample at the excitation wavelength not to exceed

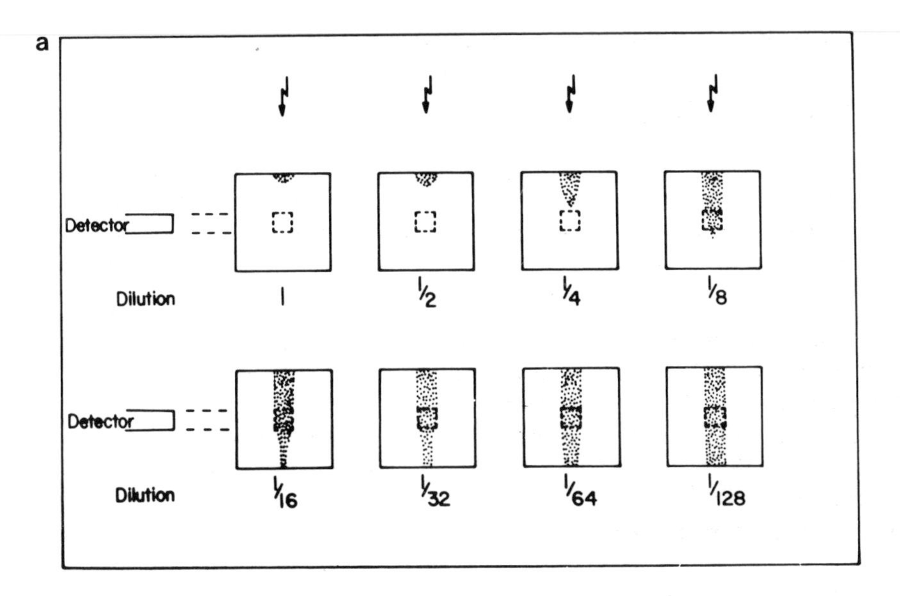

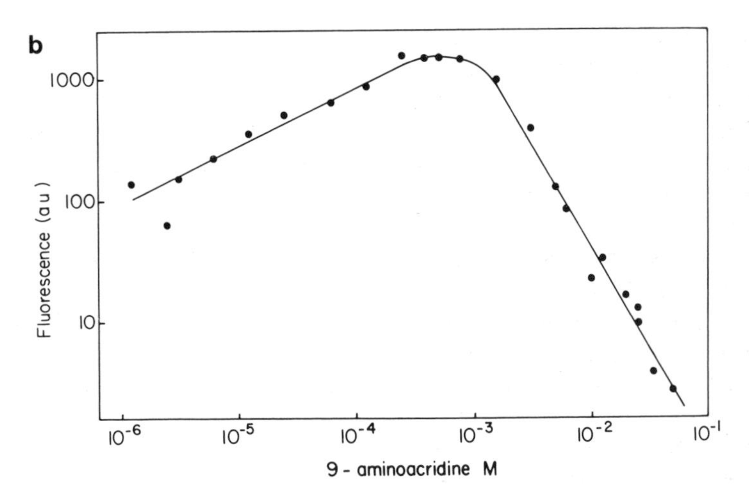

Figure 9. Fluorescence emission as a function of fluorophore concentration (**a**) 1 mM 9-aminoacridine in distilled water was placed in a quartz cuvette (all four faces polished, as is usual for fluorescence measurements) in a Perkin Elmer MPF 44A fluorescence spectrophotometer. The excitation wavelength was set to 410 nm and the slit width to 6 nm. The cuvette was observed with the sample compartment and excitation shutter open and the emission shutter closed; for best effect it is preferable to switch off the room lights. The blue fluorescence of 9-aminoacridine arose from the stippled area as indicated. Note how the machine's photodetector monitors fluorescence emission arising only within the central area of the cuvette as indicated by the inner dashed box. As the concentration of 9-aminoacridine was reduced by serial dilution, the blue fluorescence is detected from progressively more of the excitation light path (the direction of incidence is indicated by the arrow). (**b**) 9-aminoacridine dissolved in double distilled water was excited at 400 nm and the fluorescence emission at 456 nm (●) recorded with a Perkin Elmer MPF 44A fluorescence spectrophotometer with excitation and emission slits of 2 nm. Fluorescence was recorded using the 'front face' system with the fluorescent solution contained in a 2 mm pathlength absorbance cuvette whose polished face was set at 60° with respect to the excitation beam. Even with this geometry fluorescence intensity is a non-linear function of fluorophore concentration.

0.05/cm for analytical fluorescence measurements.

Remember that it is not only the fluorophore that may absorb the excitation light and that it is the *overall* absorbance that must be kept low to avoid inner filter effects. If it proves impossible to reduce the absorbance of the sample, inner filter effects can also be obviated by altering the geometry of the fluorescence experiment: fluorescence from the surface of highly absorbing material is much less sensitive to the effect because the excitation does not have to penetrate very far in order for a fluorescence signal to be recorded. Most fluorescence spectrometers provide a 'front face' attachment for recording signals from highly absorbing media. In using such attachments the specularly reflected (mirror-like) light must be excluded from the detector either by appropriate blocking filters or by choosing an angle at which to detect fluorescence that excludes the specular reflection; for example by setting the fluorescent surface at an angle of 60° to the excitation beam.

3.3 Measurement of Fluorescence Values

The general procedures for making fluorescence measurements are exactly the same as those for making absorption measurements (Section 2.3). It is particularly important to allow the light sources to stabilise before making fluorescence measurements because of the direct relationship between excitation and emission intensity. The following sequence is then performed.

(i) Select the appropriate excitation and emission wavelengths using either filters or monochromators. In the case of interference filters check that they are inserted in the correct orientation with respect to the light beam (this is sometimes indicated by an arrow engraved on the edge of the filter).

(ii) Choose the correct slitwidth for both emission and excitation.

(iii) Check the 0% fluorescence level. This is usually done by interposing a shutter between the sample and the detector. As there is no absolute scale of fluorescence it is very important to set the zero level correctly as this is the only absolute reference point available. It is advisable to check that the zero level does not change when the amplification (gain) of the detector is altered. The importance of the zero level is so crucial that we recommend it is checked periodically during a sequence of measurements as undetected 'drifts' are common and a source of significant inaccuracy.

(iv) Read the fluorescence of the reference (blank), sample and standards. It is useful to use the same cuvette throughout such a series so that errors due to incorrect matching of cuvettes are minimised. However, it is essential to rinse the cuvette thoroughly between readings to remove all trace of the preceding solution.

(v) It is useful on occasions to verify directly that the material being studied is indeed fluorescent (see for example *Figure 9*). The faintness of fluorescence may require the direct observation to be done with 'dark-adapted' eyes.

3.4 Common Problems of Fluorescence Measurements

3.4.1 *Reference Material is as Fluorescent as the Sample*

Fluorescence is particularly sensitive to contaminating substances. Thus, for example, the plasticisers present in plastic storage bottles are fluorescent, and when they leach into the water they may be present in both sample and reference solutions.

Chemicals of the highest purity may still contain traces of fluorescent contaminants which can be detected only by a careful and systematic study of all the components in the fluorescence cuvette.

The signal detected may not be fluorescence at all. Raman scattering, and Rayleigh scattering, both provide signals to the detector which can be mistaken for fluorescence. As these are usually properties of the solvent they occur in both sample and reference.

The cuvette may be contaminated by fluorescent material. Remember that fingerprints are fluorescent. The need for cleanliness is particularly stringent in fluorescence measurements.

3.4.2 *Fluorescence Reading is not Stable*

As in absorbance measurements (Section 2.4.2) this can arise for a variety of reasons.

(i) Fogging of the cuvette when the contents are much colder than the ambient temperature.
(ii) Drops of liquid on the external faces of the cuvette.
(iii) Light passing through the meniscus of the sample.
(iv) Bubbles forming in the solution as it warms.

These matters of experimental technique can be easily remedied by good spectroscopic practice.

Fluorescence is strongly reduced in the presence of certain agents known as quenchers. The most troublesome of these is molecular oxygen. If oxygen is admitted to an anaerobic system it is likely that the fluorescence will decrease. This will be time dependent if the oxygen is entering only at the air–liquid interface. It is important to remember that the solubility of oxygen in organic solvents such as ethanol greatly exceeds its solubility in water so mixing reagents dissolved in different solvents may introduce quenchers in an unexpected fashion. If all solutions are equilibrated with air the oxygen may adversely affect the sensitivity of the assay but have little effect on the stability of the reading.

The bright sources used to provide adequate fluorescence excitation also promote photochemistry and photobleaching (Section 1.1), that is light-dependent destruction of the fluorophore. This is another factor that may lead to time-dependent loss of fluorescence. Regrettably photobleaching is often irreversible. To overcome photobleaching it is necessary to reduce the intensity and energy of excitation by reducing the excitation slit and by choosing a wavelength to the red end of the excitation spectrum. If photobleaching is severe it may be necessary only to admit excitation light for a very restricted period, for example by using an instrument with a pulsed light source.

3.4.3 *Sensitivity is Inadequate*

Fluorescence is an extremely sensitive technique. However, when following published protocols it may appear that the apparatus is not operating at the expected sensitivity. Manufacturers usually provide an indication of the performance of their instruments and this should be checked from time to time. Instruments which are used with u.v. radiation for extended periods may lose sensitivity as the transparency of their optical elements declines.

When the instrument is performing to specification and yet the sensitivity of a particular assay is low the most likely explanation is the presence of fluorescence quenchers in the assay. These include oxygen (see Section 3.4.2) and metal ions. The latter are often introduced by over zealous cleaning of cuvettes in chromic acid. If this is suspected the remedy is to soak the cuvettes in solutions of chelating agents such as EDTA and to switch to a cleaning procedure that avoids metal ion contamination. The handbook from the u.v. visible spectrometry group on fluorescence gives a good guide to cuvette cleaning procedures (1).

Fluorescence is much more sensitive to changes in the environment than is absorption and sensitivity may be reduced due to a simple environmental factor such as temperature. In general fluorescence increases as temperature decreases because fluorescence emission is relatively temperature insensitive in comparison with environmental quenching processes. Thus on warm days intensities may be noticeably lower than on cool days. Many fluorimeters allow for temperature control of the sample chamber and for day-to-day reproducibility this is an important facility.

3.5 **Specialised Applications of Fluorescence Measurements**

3.5.1 *Fluorescence Polarisation*

The sensitivity of fluorescence to environmental factors has led to the development of a number of specialised applications of the technique. Of these the most widely available is the use of polarised light to yield information concerning molecular motion.

When an array of fluorophores is excited with plane polarised light those molecules with absorption moments parallel to the plane of polarisation are preferentially excited. The fluorescence emission will also be polarised provided that the chromophore remains stationary during the excited state. Movement of the excited molecules will diminish the polarisation of the emitted light. Thus the polarisation of the fluorescence emission depends on the polarisation of the excitation, the lifetime of the excited state and the mobility of the fluorophore. Small molecules (molecular weight <1000) with fluorescence lifetimes of around 10 ns (fluorescence lifetime is the time taken for $1/e$ of fluorophores excited by a single flash to return to the ground state by fluorescence emission) lose all their fluorescence polarisation when they tumble freely but retain polarisation when they are bound to macromolecules or membranes. Furthermore molecules which depolarise completely in aqueous media do not do so when they are suspended in more viscous environments. This property has led to the development of techniques to monitor the binding of haptens to antibody molecules and the microviscosity of biological membranes using fluorescent probes (6). Fluorescence polarisation may also be useful for looking at appropriately labelled macromolecules.

Polarisation of fluorescence (p) is defined as:

$$p = (V_V - H_V)/(V_V + H_V) \qquad \text{Equation 13}$$

where the subscripts refer to the direction of polarisation of the exciting light and the capitals to the emitted light: V and H refer to the vertical and horizontal directions defined with respect to the plane of propagation of light through a 90° geometry. Sometimes fluorescence anisotropy (a) defined as:

$$a = (V_V - H_V)/(V_V + 2H_V) \qquad \text{Equation 14}$$

is used instead of fluorescence polarisation.

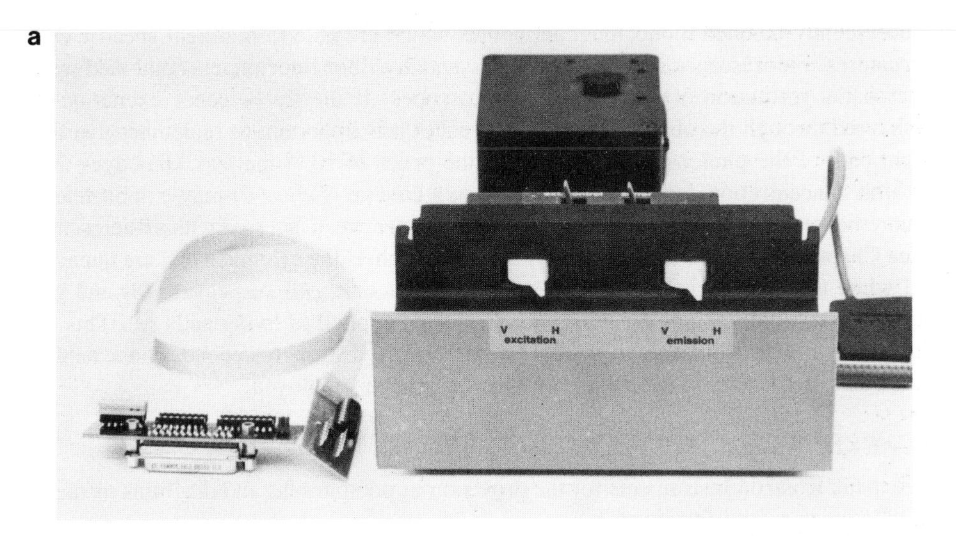

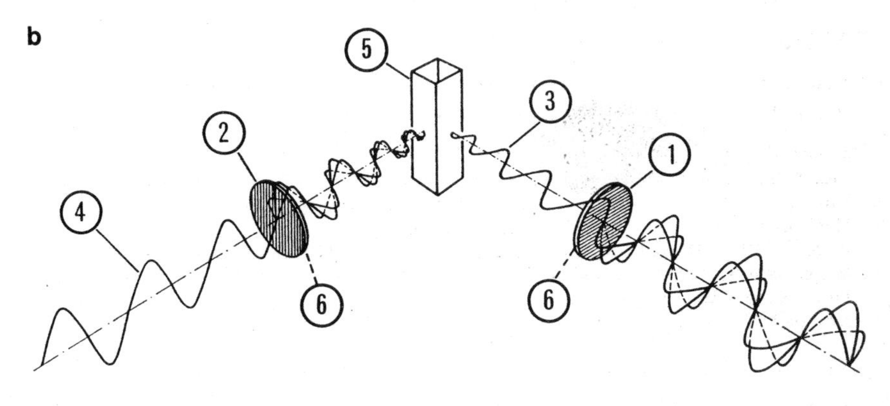

Figure 10. Fluorescence polarisation. (**a**) Automatic fluorescence polarisation accessory required for a KONTRON SFM 25 spectrofluorimeter. (**b**) Principle of the accessory illustrated above. 1: polarising filter in the excitation beam; 2: polarising filter in the emission beam; 3: polarised excitation; 4: partially depolarised emission; 5: sample in cuvettes; 6: motor to position the polarising filters (vertical and horizontal).

Figure 10 illustrates the arrangement of polarisers necessary to measure fluorescence polarisation or anisotropy in a typical polarisation accessory. It is usual in polarisation measurements to correct for differences in the transmission of the polarisers by measuring the vertical and horizontal components of the emission with the excitation polariser in both the vertical and the horizontal positions:

$$p = [V_V - H_V(V_H/H_H)]/[V_V + H_V(V_H/H_H)] \qquad \text{Equation 15}$$

Thus four separate measurements are needed to calculate fluorescence polarisation of a sample. Apparatus for the automatic calculation and presentation of fluorescence polarisation has been described (7).

3.5.2 *Fluorescence Microscopy*

Fluorescence microscopes are increasingly used in modern biology with the advent of fluorescently-labelled monoclonal antibodies which are able to highlight specific cell structures. Fluorescence microscopes are essentially filter fluorimeters combined with the spatial resolution of good optical microscopes. If the fluorescence excitation is delivered through the objective (epifluorescence) it is important to remember that the light path of the photons is very short, of the order of 10^{-6} metres. Thus dyes that exhibit concentration dependent quenching in a cuvette (*Figure 9*) may exhibit linear fluorescence with concentration over a very wide range when assessed with a microscope (see Chapter 5). Indeed a number of vital stains, such as the cyanine dyes, are quenched when they are accumulated by a mitochondrial or a cell suspension (8) and yet specifically highlight cellular mitrochondria when applied to living cells (9). This illustrates most clearly the fact that fluorescence observed strongly depends on the nature (particularly the optical geometry) of the fluorimeter employed.

4. ACKNOWLEDGEMENTS

We thank Kontron Instruments for the provision of photographs and diagrams of spectrophotometers and fluorimeters.

5. REFERENCES

1. Miller,J.N. (1981) *Standards in Fluorescence Spectrometry*, Chapman Hall, London.
2. Burgess,C. and Knowles,A. (1981) *Standards in Absorption Spectrometry*, Chapman Hall, London.
3. Rickwood,D. and Hames,B.D. (eds.), (1982) *Gel electrophoresis of nucleic acids — A Practical Approach,* IRL Press, Oxford and Washington D.C.
4. Chance,B., Legallais,V., Sorge,J. and Graham,N. (1975) *Anal. Biochem.*, **66**, 498.
5. Parker,C.A. (1968) *Photoluminescence of Solutions,* Elsevier, Amsterdam.
6. Shinitzky,M. and Inbar,M. (1974) *J. Mol. Biol.*, **85**, 603.
7. Bashford, C.L., Morgan,C.G. and Radda,G.K. (1976) *Biochim. Biophys. Acta*, **426**, 157.
8. Waggoner,A.S. (1979) in *Methods in Enzymology*, Vol.**55** Fleischer,S. and Packer,L. (eds.), Academic Press, NY p. 689.
9. Johnson,L.V., Walsh,M.L., Bockus,B.J. and Chen,L.B. (1981) *J. Cell Biol.*, **88**, 526.

CHAPTER 2

Spectra

ROBERT K.POOLE and C.LINDSAY BASHFORD

1. INTRODUCTION

Spectrophotometry and spectrofluorimetry are concerned with the ultraviolet (185–400 nm), visible (400–700 nm) and infra-red (700–15 000 nm) regions of the electromagnetic radiation spectrum. Biochemical applications, however, are generally restricted to the region between 200 and 900 nm (*Figure 1*), making u.v.-visible spectrophotometers and fluorimeters the most useful photometric instruments in biochemical laboratories.

Scanning spectrophotometers measure the absorption of light by a sample, as the wavelength of the incident light is continuously varied by a monochromator. The resulting absorbance spectrum is conventionally presented as the light absorbed (A) on the ordinate versus the wavelength (λ, in nm or formerly mμ) on the abscissa. Such spectra are widely used in the identification and quantification of compounds and can give structural information concerning the chromophore. Fluorescence is the emission of light by a sample that absorbs light (of higher energy) at a lower wavelength. Spectra are presented: (i) as fluorescence emission spectra, i.e. the wavelength-dependent

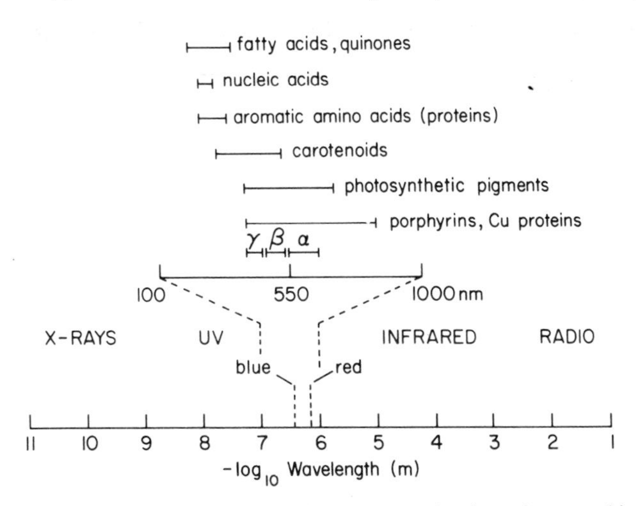

Figure 1. Schematic diagram of the electromagnetic spectrum and regions of common biochemical interest. Note (i) that the bottom scale is not linear, (ii) boundaries between adjacent regions are arbitrary and (iii) the classes of compounds shown as examples will have spectral characteristics in addition to those most commonly utilised and shown. 'Blue' and 'red' refer to the light absorbed; since the human eye, however, can detect only reflected (or transmitted) light, a solution that appears blue actually absorbs in the red at 600–700 nm.

variation in fluorescence intensity of the sample when excited at a pre-determined and fixed (lower) wavelength; (ii) as fluorescence excitation spectra, i.e. the variation in intensity of sample fluorescence measured at a pre-determined and fixed wavelength when the sample is irradiated with different excitation wavelengths. Examples of these various types of spectrum will be presented as we attempt to describe the acquisition and use of spectra.

Emphasis must be placed on the choice of operating conditions for both types of instruments; technical information on spectrophotometers and spectrofluorimeters will be included only where it assists this aim. For more general accounts of absorption and fluorescence spectra, the reader is referred to Bell and Hall (1) and Brown (2). The techniques are illustrated largely by reference to electron transport pigments, which have been the subjects of intensive and elegant spectrophotometric studies and are a research interest of the authors. Spectra that are obtained not by automatic wavelength scanning but by point-by-point measurements at discrete wavelengths, such as photochemical action spectra and kinetic difference spectra, are treated in Chapter 7.

We wish to emphasise that this chapter is in no way meant to replace the instrument manuals provided by the spectrophotometer manufacturers. (In any event, this would be an impossible task, given the fundamentally different modes of operation and the versatility of spectral measurements.) We hope, however, that the information and hints described will provide the newcomer with sufficient confidence (and a little wisdom) to attempt the measurement of absorption and fluorescence spectra of his own material and to make the trials of the types illustrated in some of our figures.

2. TECHNICAL ASPECTS

2.1 **Optical Layout**

The essential features of a simple scanning spectrophotometer are:

(i) one or more light sources emitting a continuous range of frequencies in the u.v.-visible range;
(ii) a monochromator based on a prism or grating, driven to scan the desired wavelength range;
(iii) a sample or cuvette holder;
(iv) a detector, generally a photomultiplier although silicon diode detectors are increasingly common in modern instruments;
(v) the hardware required to coordinate wavelength scanning with absorbance measurements and to plot the resulting spectrum.

Since the lamp intensity varies with wavelength, automatic scanning instruments must also compensate for this, generally by controlling the slitwidth or, in most modern instruments, by automatic gain control of the signal amplifiers. These features are common to all scanning spectrophotometers and there are also striking similarities in the accessories and operating variables offered. However, there are also fundamental areas of distinction, which are not always apparent from a superficial examination of the instrument, nor sometimes even from the manufacturer's literature. When confronted with an unfamiliar spectrophotometer or the desire to record absorption spectra in an instrument used previously only for 'single wavelength' measurements, the first priority should be to establish which modes of operation are available and which, if any, of these are suited to the application in hand. Many instrument manuals include diagrams of the optical layouts (see Chapter 1), which should be consulted in association with the descriptions that follow. Unfortunately, there is a long-standing confusion among users (and sometimes sales representatives) regarding the nomenclature of spectrophotometers. With few exceptions (notably the stable, single beam DU7 series recently introduced by Beckman), instruments designed for wavelength-scanning employ either the split beam or dual-wavelength principles.

2.1.1 *Split Beam*

The authors and certain manufacturers (e.g. Aminco) use the term 'split beam' to describe the time-sharing (or splitting) of one monochromatic beam of light (which may be varied with time) between two cuvettes, a sample (or test position) and a reference. Indeed the presence of two cuvette positions, each in a separate light beam, is characteristic of this type of instrument. 'Double beam' sometimes describes the same configuration, which has two notable applications.

(i) In comparing a sample and reference at each wavelength, the split beam spectrophotometer produces a 'difference' spectrum in which those features common to sample and reference are cancelled out. [Note, however, that where the contents of the reference cuvette are known to have no significant absorbance features in the region scanned, a spectrum recorded in this 'difference' mode is frequently called an 'absolute' spectrum. In the Hitachi 557, difference spectra can be obtained by subtraction of two such 'absolute' spectra (the so-called 'memory method').] This allows, for example, the elimination of absorbance features of the solvent (if it is included in both cuvettes) or the non-specific, light-scattering of two turbid suspensions. *Figure 2* illustrates an attempt to record the absolute spectrum of yeast cells in a suspension of modest turbidity, and the use of the difference method to cancel (or 'buck-out') the turbidity and obtain, after suitable scale expansion, a useful difference spectrum of the chromophores. The importance of the detector geometry relative to such optically unfavourable samples was discussed in Chapter 1.

(ii) Sometimes the information content of a spectrum is greater when it is the difference between two chemically differentiated forms of the sample. For example, in the reduced-*minus*-oxidised spectrum of respiratory chain components (*Figure 2*), the scanning beam is time-shared between a reduced sample (traditionally in the front or sample position) and an oxidised sample (in the reference position). The absorption bands of the reduced cytochromes appear as peaks, whilst components having more characteristic bands in the oxidised (e.g. flavoproteins), or oxygenated, forms (e.g. certain oxidases) will appear as troughs. Other examples occur in the study of ligand-bound forms and photodissociation spectra (Section 4).

2.1.2 *Dual-wavelength*

In principle, kinetic data could be achieved by repetitive wavelength-scanning in a split beam apparatus, followed by re-plotting, as a function of time, the absorbance changes at desired wavelengths. In practice, however, kinetic measurements that require

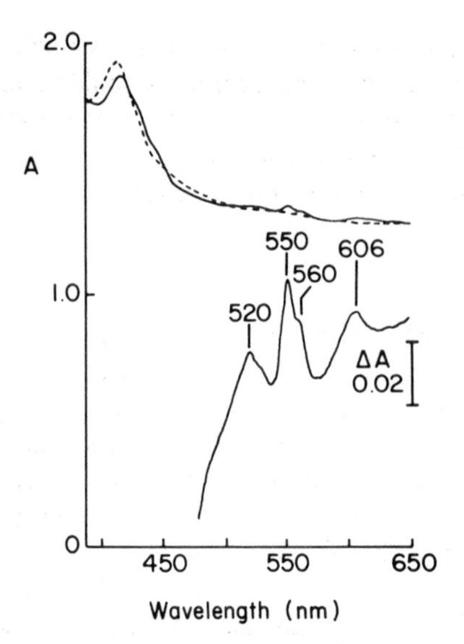

Figure 2. Attempted 'single beam' or absolute analysis of cytochromes in a turbid suspension. A suspension of *Saccharomyces cerevisiae* in 50 mM phosphate buffer (pH 7.0) was adjusted to give an A_{390} of about 1.8. The dashed line is the spectrum of the native suspension recorded with an SP1700 spectrometer and using buffer in the reference cuvette. The upper, solid line is the spectrum of a similar yeast sample reduced with $Na_2S_2O_4$. Note that the absorbance due to cytochrome c (+ c_1) at 550 nm is only about 2% of the 'apparent absorbance' (largely light scattering) of the turbid suspension. Below is shown, at higher gain, the reduced *minus* oxidised difference spectrum of the same samples. The scan rate was 4 nm/sec and the spectral band width 2 nm. Reproduced with permission from ref. 11.

compensation for high turbidity of the sample are made using a dual-wavelength (or two wavelength) apparatus in which a single sample is alternately illuminated with two wavelengths (λ_1, λ_2). The reference wavelength (λ_2) will preferably be at an isosbestic point (see below) so that the absorbance at that wavelength will remain relatively constant and serve as a measure of the sample turbidity. The difference ($A_{\lambda_1} - A_{\lambda_2}$ often written as $A_{\lambda_1 - \lambda_2}$) is recorded and plotted. If λ_1 is now scanned, we have the basis of a dual-wavelength (or two wavelength) scanning spectrophotometer, a principle offered in an increasing number of commercial instruments. Only one cuvette is used and so difference spectra can be obtained only by recording successively spectra of two samples and computing the difference.

2.1.3 *Single Beam*

With the decline in the cost of microprocessors and microcomputers, there has been a resurgence of interest in single beam instruments (i.e. using one cuvette and one, unchopped light beam) for wavelength scanning. The spectrum of the reference sample is stored in instrument memory and subtracted electronically from subsequent scans.

2.1.4 *Making a Choice*

Some commercial instruments (e.g. the Hitachi/Perkin-Elmer 557 and Aminco DW2) offer a choice of operating modes suitable for wavelength scanning, most commonly split beam or dual-wavelength. Beckman now market rapid scanning, single beam instruments (the DU7 series) which have sufficiently versatile software for them to provide useful spectra from samples which formerly could be monitored only using split beam or dual-wavelength instruments. A choice will have to be made based on the constraints of the sample (e.g. amount available) and experience. It can be instructive to run the same sample in the various modes offered.

2.2 **Baselines**

A baseline is the wavelength-dependent difference in absorbance either (i) between two cuvettes in a split beam apparatus when their contents are thought to be identical or (ii) between two scans of one cuvette in a dual-wavelength spectrophotometer over a period when the contents are thought not to have changed. Any irregularities in a baseline will be included in subsequent difference spectra and may be superimposed on the spectral regions of interest. In extreme cases, baseline irregularities may be mistaken for spectral peaks and troughs, and a steeply sloping baseline can alter the apparent position of an absorption peak. Therefore, baseline flatness should always be checked before recording a difference spectrum and either presented with the latter or, at least, reported.

The following additional measures can be taken to improve baseline flatness.

2.2.1 *Electronic Baseline Correction Facilities*

In the Aminco DW2, for example, the baseline can be adjusted by a series of either 11 or 31 potentiometers that effectively add to, or subtract from, the signal over the selected wavelength range. Over 300 nm, the series of 11 adjustments will allow a correction every 30 nm. The system is quite effective, but tedious, and adjustments must be re-made when the sample or its concentration is changed. A more convenient aid is exemplified by the Hitachi 557. Here, any irregularities are 'memorised' by the instrument in an initial scan (normally not plotted) of the cuvette(s) and then subtracted from subsequent scans. Remember, however, that in this case, any dissimilarities in the cuvettes are essentially hidden.

2.2.2 *Sample Preparation*

Give extra care to cleanliness and matching of the two samples, the proper alignment of the cuvette(s) and the identicality of the cuvettes' contents. Where both test and reference cuvettes contain samples treated in some way, such as is the case in the recording of the baseline (reduced versus reduced) for a CO difference spectrum (*Figure 3*), it is generally preferable to split the sample between the cuvettes after (in this example) reduction.

2.2.3 *Detector Organisation*

Try rotating the detector, if its housing allows, to reduce the effect of non-uniformity in the photocathode sensitivity. A quartz diffuser plate inserted in front of the detector scatters radiation over a wider area of the photocathode and may achieve a similar result.

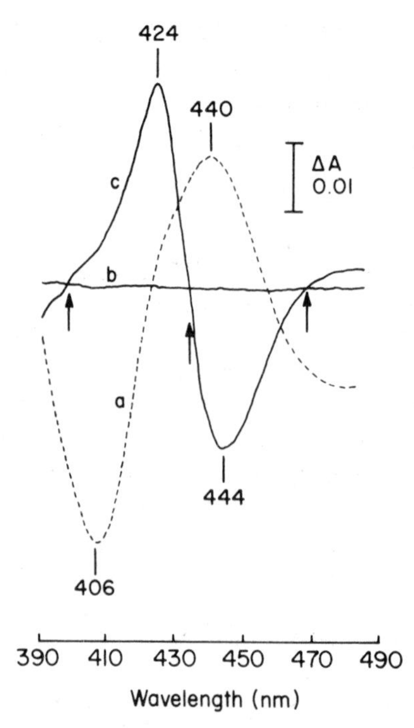

Figure 3. Difference spectra of the purified 'haemoprotein b^{-590}' from *Escherichia coli*. The dithionite-reduced *minus* native (oxidised) difference spectrum is shown in (**a**); this was obtained in the split beam mode after establishing an oxidised *minus* oxidised baseline (not shown), using the baseline correction facility of the Hitachi 557 spectrophotometer. The baseline exhibited no irregularities. A second, reduced *minus* reduced baseline was obtained in the same way (**b**), before bubbling the contents of the sample ('front') cuvette with CO for 2 min (**c**). Note the loss of the 440-nm band of the reduced haemoprotein on bubbling with CO and the appearance of a ferrous carbonmonoxy form at 424 nm. The isosbestic points arrowed are at 398, 434 and 468 nm. The spectra were obtained in a pathlength of 10 mm, with a scan speed of 2 nm/sec, a spectral band width of 2 nm and the 'auto' response setting. Unpublished experiment of R.K.Poole and C.A.Appleby.

2.2.4 *Manual Correction*

In the absence of the above facilities, the uncorrected baseline can be recorded and subtracted from the difference spectrum manually at, say, 2 nm intervals. The difference spectrum can be assumed to cross the baseline at a known isosbestic point.

2.3 **Isosbestic Points**

An isosbestic point occurs at a wavelength where the absorptions of two light-absorbing forms are equal (Greek *iso*, equal; *sbestos*, extinguished) and may be seen, for example, where a difference spectrum crosses an appropriate baseline. The isosbestic point is useful in both quantitative and qualitative work. Where a clear isosbestic point occurs during the course of a reaction (*Figure 4*), it is often taken as evidence that only two species are involved but it is conceivable that a third species may simply have no absorption at this wavelength. However, the *lack* of an isosbestic point does indicate the presence of a third component, unless one of only two species exhibits a deviation from the Beer–Lambert Law. The isosbestic point is more generally used as a reference

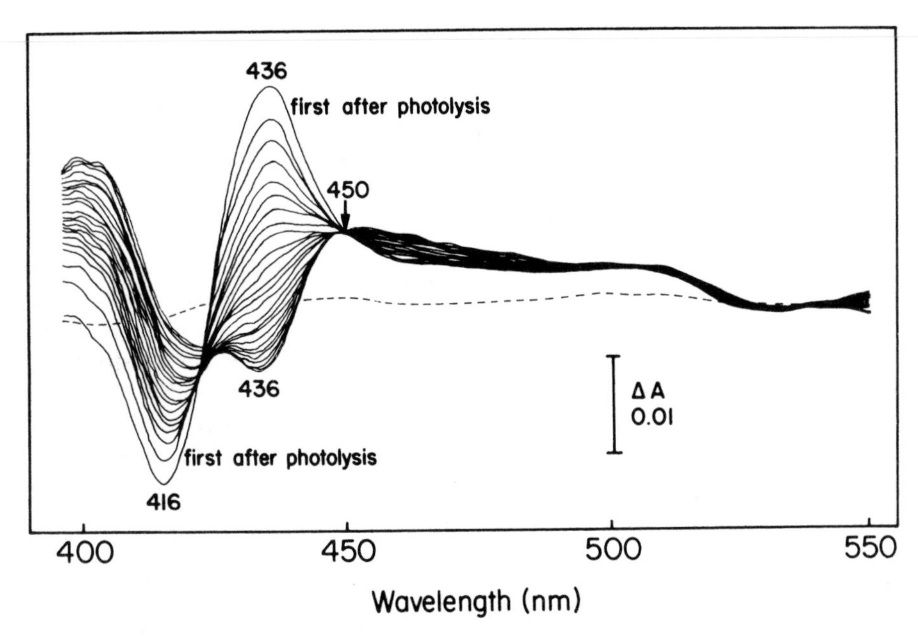

Figure 4. Example of an isosbestic point in the reaction with oxygen of cytochromes in *E. coli* at −98°C. The spectrum of a suspension of reduced, CO-liganded cells was recorded and stored in the digital memory of a dual-wavelength scanning spectrophotometer, with a reference wavelength of 575 nm. Subsequent scans are difference spectra with the stored spectrum subtracted. The first scan before photolysis yielded the baseline shown (dashed). The reaction was then initiated by photolysis and followed by repetitive scanning. For further details, see ref. 12. Photolysis gives the 436 nm peak of the reduced cytochrome *o*; during the time interval shown, the combination of this species with oxygen gives a clear isosbestic point at 450 nm. Between 416 and 436 nm, however, an isosbestic point is not seen, implicating the presence of a third component. Reproduced with permission from ref. 12.

wavelength to which the absorbance at a nearby wavelength may be referred in (i) setting up a dual-wavelength spectrophotometer or (ii) measuring from an absorption spectrum the concentration of a component. Less frequently, it is a wavelength suitable for quantification of the total amount of the two species present (*Figure 4*). Finally, in the presentation of many stacked spectra, quoting the isosbestic point(s) eliminates the need to draw baselines for each spectrum.

2.4 Wavelength and Absorbance Calibration

Factors that can affect the apparent position of an absorption maximum are described in Section 4. Even when operating conditions are properly chosen, the possibility remains that the spectrophotometer generates peaks that are a nanometre or so from the expected position. Wavelength accuracy can be checked using the bright line spectrum (λ_{max} = 656.1 nm) of the deuterium light source (as recommended for the Hitachi 557) or standards for checking the wavelength calibration are available. Most commonly, these are rare earth oxides or a mixture of oxides such as holmium and/or didymium; such 'filters' should be scanned with air as the reference or blank. Holmium has nine useful maxima between 241.5 and 637.5 nm and didymium five between 573 and 803 nm (see ref. 3, and the manufacturers' charts). A quick and easy check of both wavelength calibration and accuracy of absorbance measurement is to use a 0.2 mM solution of potassium chromate in 0.05 M KOH. Two broad absorption bands at 275 nm and 375 nm should be seen with molar absorptivities, respectively, of 3680 and 4820 in a 1 cm cell (4). It is useful to run calibration spectra periodically, dating the scans, to check for degradation of accuracy and calibration.

2.5 **Stray Light**

This is light received at the detector that is not anticipated in the spectral band isolated by the monochromator. Stray light is usually 'white' (i.e. having the same composition as the source or ambient illumination) and arises from (i) the monochromator itself and (ii) light leaks in the sample and detector region. The first should be minimised by good monochromator design and filters and the second by proper construction of the instrument. Consequently, stray light should not be a problem with a high quality spectrophotometer, where linearity to an absorbance of 2 or 3 may be expected, but it is wise to determine the range in which the measured absorbance in the selected wavelength range is directly proportional to sample concentration, i.e. the range where the Beer–Lambert Law applies. The effect of various degrees of stray light on this relationship are shown by Wood (5) and Gratzer and Beaver (4), who also outline methods for estimation of stray light. Some manufacturers (e.g. Gilford) provide neutral density filters with known transmission characteristics which can be used to check for stray light and electronic errors in the absorbance output. Kodak Wratten neutral density filters, which are cheap and easily obtainable, can be used for the same purpose. Thus insertion of a filter of neutral density 1 into the measuring beam should yield an absorbance reading of 1, insertion of two such filters should yield an absorbance of 2 and so on (i.e. neutral density should be strictly additive).

2.6 **Fluorimeters**

Fluorescence measurements are almost universally made in a 'single beam' mode and it is normal for instruments to provide both emission and excitation spectra which depend, to a greater or lesser extent, on the characteristics of the fluorimeter as well as those of the sample. This arises, in part, because there is no absolute scale of fluorescence intensity (i.e. there is no equivalent to the Beer–Lambert Law in fluorescence) and fluorescence spectrometers register 'relative fluorescence', a parameter which may vary significantly from apparatus to apparatus. This absence of standardisation is not a serious handicap when measurements made with a specific apparatus are compared with each other, particularly if care is taken to establish 'standard operating conditions' for the device. With the increasing use of fluorescence techniques it is to be hoped that more laboratories will published fluorescence spectra 'corrected' for instrumental variables. The procedures necessary to obtain 'corrected' spectra can be found in references 6 and 7. 'Corrected' fluorescence excitation spectra are corrected for changes in lamp output and should correspond exactly to the absolute absorbance spectrum of the fluorescent material. In some fluorimeters, such as the Perkin-Elmer MPF 44, the excitation beam is split such that most falls on the sample but a small fraction is diverted to a photodetector which regulates the signal amplifier gain to provide a constant output in the face of fluctuations in excitation intensity. Excitation spectra acquired in this so-called 'ratio-mode' approximate more closely to the 'corrected' spectrum than do spectra acquired at constant gain (the so-called 'energy mode'). 'Corrected' emission spectra are corrected for the non-linearity of the photodetector and are usually obtained, indirectly, by comparison with standard fluorophores. Although investigators are strongly encouraged to report corrected fluorescence spectra it remains the case that in many biochemical applications uncorrected spectra are usually sufficient for correct conclusions to be drawn from particular experiments.

3. CHOICE OF OPERATING CONDITIONS AND THEIR INFLUENCE ON SPECTRAL 'QUALITY'

The choice (if available) of optical operating mode (single beam, split beam or dual-wavelength scanning) was discussed above. The factors that follow must be considered irrespective of which mode is chosen. Additional advice on 'good spectroscopic practice' can be found in the monographs published by the u.v. spectrometry group (7,8).

3.1 Wavelength Range and Light Source

In most cases this decision will be made on the basis of published optical properties of the sample in question; *Figure 1* summarises the wavelength ranges in which compounds of biochemical interest are frequently studied and Morton (9) provides a very comprehensive collection of spectral data. The lamps in most common use are the tungsten–iodide lamp for the near-u.v. (say >330 nm), visible and near-infra-red regions and the deuterium arc for other u.v. use (180–380 nm). In all cases the manufacturers' recommendations should be followed.

Fluorescence spectrometers usually employ mercury or xenon arcs to provide adequate excitation intensity. The emission from a mercury arc is confined to relatively few bands (the continuum emission level is low) which provide excellent excitation intensity provided that the sample absorbs light in this region; because of the intensity of the mercury lines such sources are *not* suitable for recording excitation spectra. Xenon arcs provide much more continuous intensity and are the preferred source for scanning fluorescence spectrometers. However, investigators should be aware that xenon arcs do provide 'bright' lines in the u.v.-visible region which may occasionally interfere with the acquisition of reliable excitation spectra. The selection of a suitable emission wavelength at which to monitor an excitation spectrum, and conversely the choice of suitable excitation wavelength for emission spectra, is not necessarily straightforward. In many instances the excitation and emission spectra overlap and if the excitation (or emission) maximum were selected simple light scattering would severely distort parts of the emission (or excitation) spectrum. However, the form (but not the intensity) of both emission and excitation spectra are usually independent of the excitation or emission wavelength, respectively. Thus it is common to select an excitation wavelength lower than the maximum excitation wavelength in order to record an entire emission spectrum; conversely an emission wavelength longer than the maximum may be needed to monitor the excitation spectrum.

3.2 Effect of Spectral Bandwidth

Perhaps more than for any other operating variable, the consequences of inappropriately chosen slit width, and thus spectral band width, are misunderstood. The resolution of

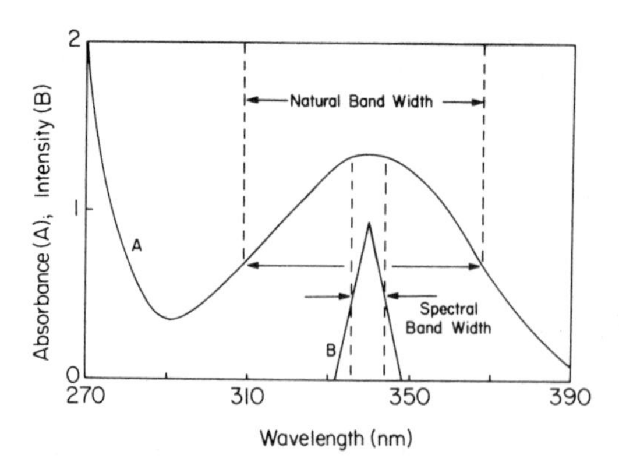

Figure 5. The relationship between natural and spectral band widths. The natural band width shown is for NADH (58 nm, **curve A**). **Curve B** is a schematic representation of the spectral bandwidth of the monochromator exit beam. Reproduced with permission from ref. 5.

a spectrophotometer, that is its ability to distinguish between two absorption (or fluorescence) bands close together, will depend on the width of the bands under study and on the width of the spectral region passed by the monochromator(s). The former is the 'natural band width' and is defined as the width (in nm) at half the height of the sample absorption peak (*Figure 5*). It is important to note that it is *independent* of instrument band width, being an intrinsic sample characteristic. For example, the value for the absorbance band of NADH at 340 nm is 58 nm (*Figure 5*), whereas for most cytochromes at room temperature, the α-bands (those in the region 540–640 nm; *Figure 1*) have natural band widths of about 10 nm. In contrast, the spectral band width is a property of the instrument and will be either given or can be calculated. It is defined as the band of wavelengths containing the central half of the entire band of wavelengths passed by the exit slit of the monochromator (*Figure 5*). It is this slit that can generally be adjusted by the user and the adjustment will be graduated in spectral band width (nm), slit width (mm) or both. To calculate spectral band width from slit width, we need to know the reciprocal dispersion of the monochromator. This is $d\lambda/dx$, where x is the distance (in mm) in the exit slit plane. Values can generally be found in the instrument manual. A typical value might be 2 or 4 nm/mm. Thus, in the latter case, an exit slit 1 mm wide would pass light of 4 nm spectral band width.

Knowing the spectral and natural band widths with which we are working, we can calculate the error in the peak height measured by the spectrophotometer at a given slit width and natural band width (*Figure 6*). For example, use of a spectral band width of 8 nm will give rise to an error of about 22% if attempting to measure the concentration of a cytochrome with a natural band width of 10 nm, but in a measurement of NADH (natural band width 58 nm) the error will be less than 0.5%. Natural band width is temperature-dependent, so that low temperatures require the use of narrower slits. It might seem from the above that narrow slits (and thus small spectral band widths) would always be chosen. The penalty, however, is a reduction in light reaching the sample and detector with a consequent increase in 'noise'. By this, we mean any undesired recorder pen excursion (normally measured as a peak-to-peak level) superimposed on the signal of interest. For quantitative work with compounds having relatively narrow

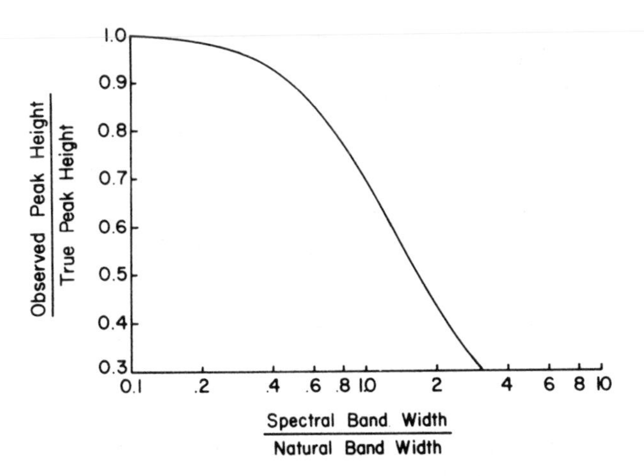

Figure 6. Graph for calculation of the error in measured peak height for a given monochromator slit width and natural band width of an absorption band. See text for examples of its use.

band widths, a compromise mut be reached between error in peak heights and an undesirable signal-to-noise ratio. Probably, the best solution is to try different spectral band widths (*Figure 7*), remembering that because, for example, of sample light-scattering and non-uniform detector response, one slit width may not be ideal for *all* regions of the spectrum. With the monochromator set not to scan, it can be useful gradually to close the slit while watching the response of the pen, until the noise level becomes unacceptable.

In fluorescence measurements, slit width can be adjusted both on the excitation and the emission sides of the system. The relationship between natural and spectral band width outlined above applies in both cases, so that for recording excitation spectra it is desirable to minimise the excitation slit width and to use a wide emission slit, whereas for recording emission spectra one should use a broad excitation band (for efficient excitation) and minimise the emission slit width. It is fortunate that in biochemical applications most of the fluorophores encountered have rather 'broad' spectra (i.e. large natural band widths) allowing quite wide instrument slits to be employed. *Figure 7* illustrates the loss of resolution of spectral features as the spectral band width approaches that of the emission lines in the fluorescence of 9-aminoacridine.

3.3 Scanning Speed and Instrument Response Time

These variables are closely related and again their selection is a matter of compromise. Selection of too fast a scan and/or too slow a response time will cause spectral distortion (*Figure 8*), although, to a certain extent, the dangers of a fast scan speed can be counteracted with a fast response time, provided that the noise so introduced is not troublesome (*Figure 9*). Slow scan rates over a long wavelength range may allow the sample to change, e.g. by cells or particles settling out of suspension). The inclusion of a viscous solute, such as glycerol, can help in these cases.

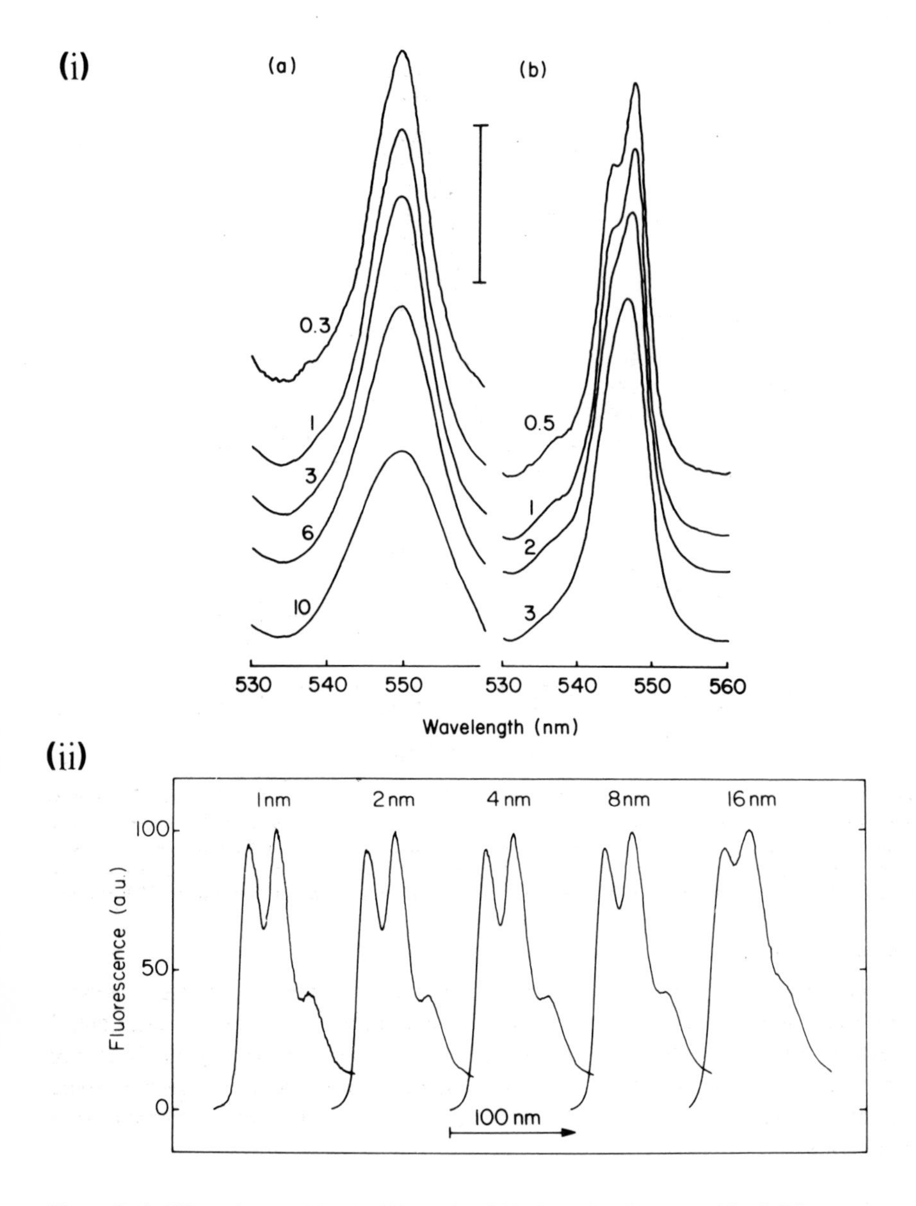

Figure 7. **(i)** Effect of spectral band width on the dithionite-reduced *minus* oxidised difference spectrum of cytochrome *c* at room temperature (**a**) and 77 K (**b**). The solution was approximately 0.25 mg cytochrome/ml in a buffer that contained 110 mM TES, 20 mM $MgCl_2$, 0.25 M sucrose and 0.25 mM EGTA, pH 7.0. The scan rate was 2 nm/sec in (**a**) and 1 nm/sec in (**b**), and in each case the medium response time (0.1 sec) of the Aminco DW2 was selected. The bar represents 0.04 *A* in (**a**) and 0.21 *A* in (**b**). Figures on the spectra are spectral band widths in nm. Note the greater tolerance to broad spectral band width in (**a**), where the natural band width is larger. **(ii)** Effect of spectral band width on the fluorescence emission spectrum of 9-aminoacridine. 10 μg of 9-aminoacridine/ml in 5 mM Hepes, pH adjusted to 7.4 with NaOH at 25°C was excited at 350 nm (10 nm slit) in a Perkin Elmer MPF 44A fluorescence spectrophotometer. Instrument response time was 0.3 sec and the scan rate 1 nm/sec, the emission slit width (spectral band width) is indicated for each spectrum. Each spectrum spans the wavelength range 400 nm (left hand edge) to 520 nm.

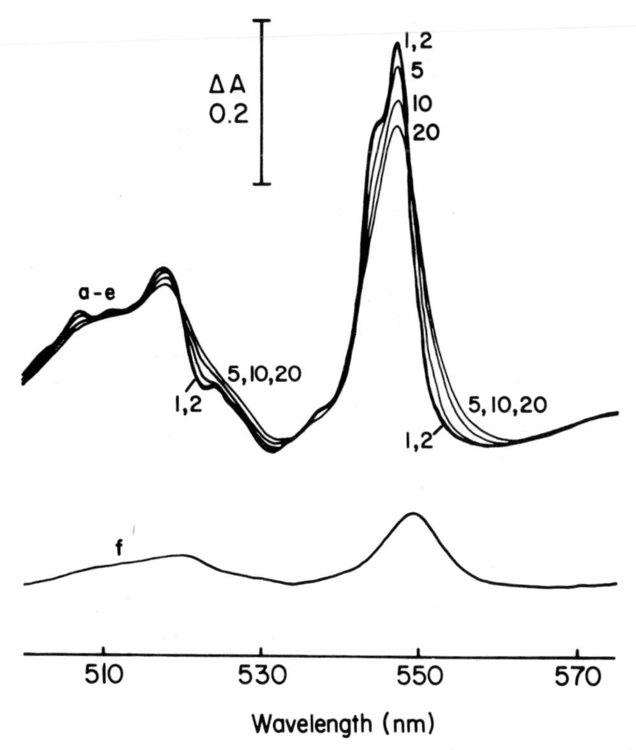

Figure 8. Effect of scan rate on the reduced *minus* oxidised difference spectrum of cytochrome *c*. The samples were the same as in *Figure 7 (i)*. The spectral band width was 1 nm, the temperature 77 K, and the response time 'medium' (see *Figure 9* for further details of this variable). The figures on spectra (**a**)−(**e**) are the scan rates in nm/sec. Spectrum (**f**) was recorded under the same conditions (scan rate 1 nm/sec), but at room temperature and shows the enhancement of peak heights, their splitting and the blue shift of the 77 K spectra.

3.4 Temperature

The temperature-controlled sample holders available in most commercial spectrophotometers, when coupled to an external circulating heater or cooler, can control temperature in the range 0−40°C. Specialised cuvettes are not necessary and the device is useful for kinetic measurements or observation of a labile sample. However, there are many advantages for wavelength scanning of being able to operate at much lower temperatures.

3.4.1 *77 K*

Spectroscopy at 77 K (liquid nitrogen temperatures) has been widely used to detect differences in the absorption spectra of closely related haemoproteins, to trap unstable intermediates and steady-states of oxidation and reduction, to slow down rapid rates of reaction and to detect and measure very low concentrations of haemoproteins (10). The main effects are a sharpening and enhancement of the absorption bands and a blue shift of 1−4 nm (*Figure 8*). The absorbance changes are due to:

(i) a true temperature dependence of the absorbance of the chromophore, resulting in narrowing, sharpening and shifting to shorter wavelengths of the bands;

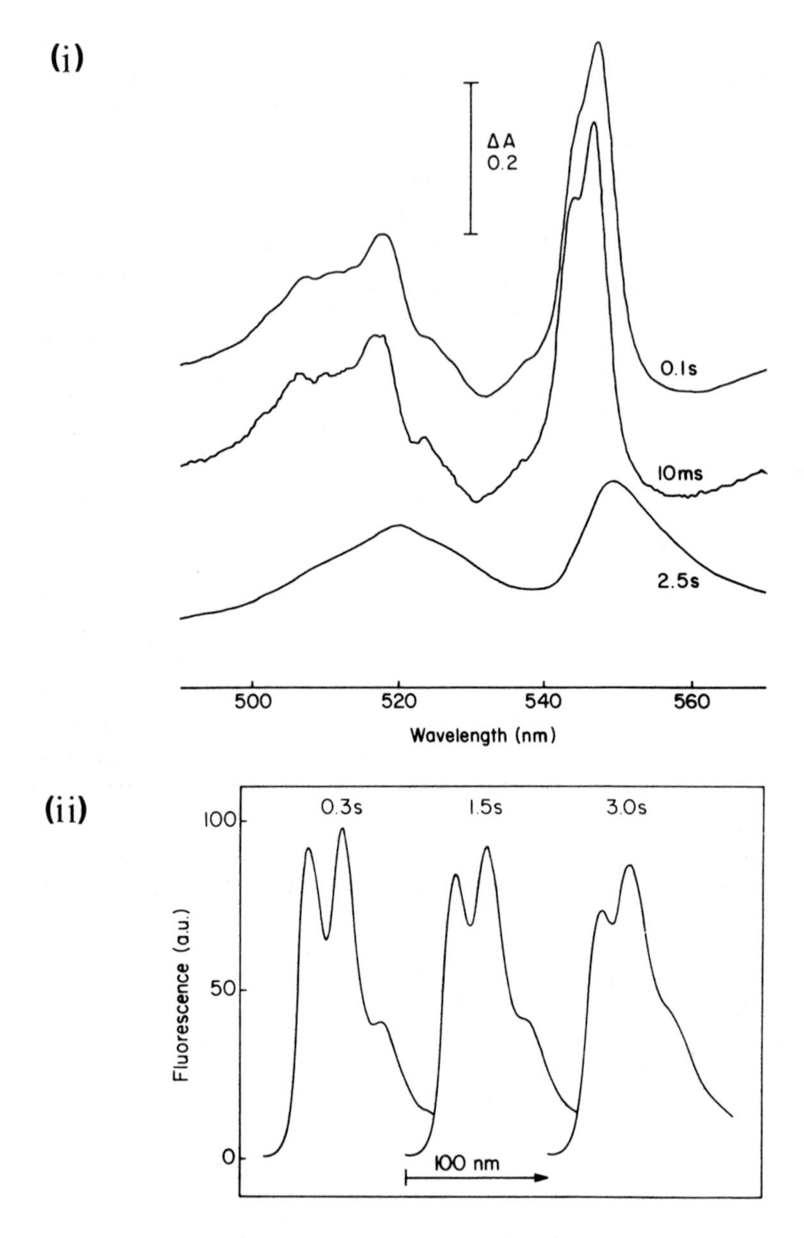

Figure 9. (**i**) Effect of instrument response time on the reduced *minus* oxidised difference spectrum of cytochrome *c*. The samples were as in *Figures 7* and *8*. The spectral band width was 0.5 nm and the scan rate relatively fast (5 nm/sec). Shown on the spectra are the response times selected, this being defined as the time required for the response to reach 63% of full scale definition. Note the severe spectral distortion evident with the two slower response times and the ability of the fastest response time to compensate for the distortion seen at this scan rate in *Figure 8*. (**ii**) Effect of instrument response time on the fluorescence emission spectrum of 9-aminoacridine. 10 μg of 9-aminoacridine/ml in 5 mM Hepes, pH adjusted to 7.4 with NaOH at 25°C was excited at 350 nm (10 nm slit) in a Perkin Elmer MPF 44A fluorescence spectrophotometer. Emission spectra were recorded using a 2 nm slit, a scan rate of 4 nm/sec and the instrument response time indicated. Each spectrum spans the wavelength range 400 nm (left hand edge) to 520 nm.

(ii) light-scattering changes in the medium, which result in an effective increase in path length by multiple internal reflections from ice crystals. [It is worth noting that enhancement effects of related origin may be observed in highly scattering suspensions (e.g. of intact cells) at room temperature.]

The enhancement effects are strongly dependent on the suspending medium and the method used to freeze it. For example, a devitrified (i.e. polycrystalline) 1.4 M sucrose solution can give a 25-fold intensification at −190°C, compared with 20°C. The presence of an organic solvent also makes the enhancement factors more reproducible than in dilute buffers (10). Glycerol has the added advantage of suppressing the pH changes that occur on freezing, as does a careful choice of buffer. The pH of Tris buffer changes dramatically (increasing) on lowering the temperature, while phosphate and acetate buffers, for example; are relatively temperature-independent.

Accessories for recording spectra at 77 K are available for many but not all spectrophotometers. The attachments are generally expensive and many workers have designed and constructed simple devices for their own instruments. References to examples are given by Jones and Poole (11), who also give further information on the attachment for an SP1700/1800 shown in *Figure 10*.

Essentially, it consists of a small Dewar flask held by clips to a brass support that slides into the space vacated by the regular two-cuvette block of the spectrophotometer. The part-silvered Dewar vessel was manufactured to order. The bottom 4 cm are silvered, such that, when in position in the spectrophotometer, the silvering stops just below the light paths. The brass cuvette holder has four screws for level location on the sometimes irregular Dewar top; this allows vertical and rotational adjustment of the cuvettes with respect to the Dewar (using the vertical threaded post and lock-nuts) and the removing of each cuvette individually, thus allowing the recording of differences between various pairs of samples. The angle between the two cuvettes equals that between the positions for 1 cm × 1 cm cuvettes in the standard cuvette holder. The path length

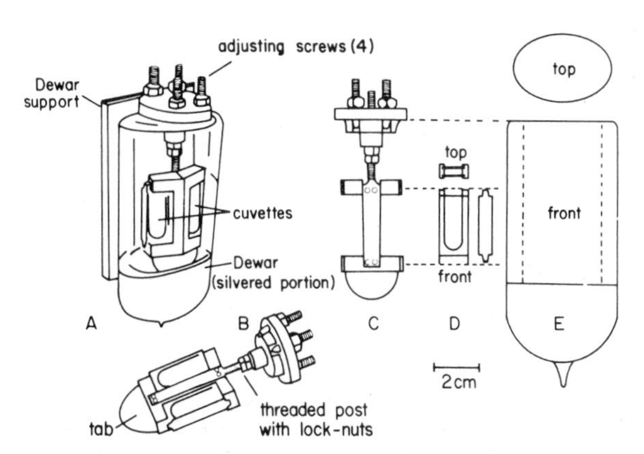

Figure 10. An attachment for the recording of difference spectra at 77 K in Unicam split-beam spectrophotometers. **A** and **B** are sketches of the complete apparatus, and the cuvette holder with cuvettes, respectively and are not to scale. **C** is a scale diagram of the cuvette holder viewed from the back (i.e. the face adjacent *in situ* to the photomultiplier). **D** shows one of the two removable cuvettes. **E** is the Dewar vessel in top and front views, and shows the extent of silvering. The scale applies to diagrams **C, D** and **E** only. Reproduced with permission from ref. 11.

of the cuvettes is only 2 mm. The small light path hastens freezing, facilitates examination of the very opaque and light scattering samples and is generally more than compensated for by the enhancement of peak heights at 77 K. In the cuvette, a brass framework supports two Perspex windows that fit tightly into grooves in the brass frame, spaced so as to give a path length of 2 mm. Although not always liquid-tight, these cuvettes are more robust, do not crack on repeated immersion in liquid nitrogen and aid rapid freezing of samples squirted into them.

In use, the Dewar is mounted in its holder in the light path and filled with liquid nitrogen to the level of the silvered margin. A moderate stream of compressed air is directed over the lower, outer surface of the Dewar to de-mist the cold glass surfaces. The cuvettes are generally positioned in their holder and filled from hypodermic syringes with needles. Using tongs, the complete assembly is immersed in a separate, large Dewar flask filled with liquid nitrogen until all bubbling stops. The cooled assembly is then located in the Dewar with the orientation of the cuvettes matching that in the standard cuvette holder. The brass semicircular tab at the bottom of the cuvette holder dips into the liquid, whilst the cuvettes are raised just clear of the surface. When the 'correct' rotational and vertical position of the cuvettes has been established, and the adjusting screws at the top of the holder have been set to hold it snugly on top of the Dewar (all adjustments being made to give the flattest baseline), the apparatus has proved to be reliable and simple in use.

Amongst commercially available designs, that for the Hitachi 557 resembles our home-made device in having the facility for individually removing the cuvettes. In the Aminco model, the 'cuvettes' are not removable from the spade-shaped structure which is inserted into the Dewar. This has the disadvantage that the sample and reference contents cannot easily be thawed and replaced individually but ensures equal freezing of both sides.

The use of low-temperature accessories in photodissociation spectra is described in Section 4.3.

3.4.2 *Other Sub-zero Temperatures*

In media, such as those containing mannitol or sucrose, which do not exhibit phase transitions in the experimental range of temperatures, the absorption peak heights and peak areas are nearly linear functions of temperature between −40°C and liquid nitrogen temperatures. Furthermore, if the sample is warmed within this temperature range, and cooled again to the original temperature, the resultant spectrum is indistinguishable from that recorded originally (10). At temperatures higher than −40°C, the absorbance intensity decreases abruptly and re-freezing does not fully restore the enhancement.

For kinetic studies at sub-zero temperatures, or where it is required to study the effect of temperatures on an absorption spectrum (e.g. in investigating haemoprotein spin-states), it is sometimes necessary to use temperatures other than 77 K. Temperatures between about 0 and −40°C can be maintained simply by circulating cooling water with ethylene glycol around a cuvette containing the ethylene glycol-supplemented sample. For temperatures between about −40°C and −140°C, as required for kinetic studies of ligand binding to cytochrome oxidases and of subsequent electron transfer, the procedure developed by Chance and co-workers (12) for their 'triple trapping' studies of mitochondrial oxidases is appropriate. Here, the temperature is maintained by a steady flow of nitrogen gas, cooled by passing through a copper coil immersed in liquid nitrogen and re-heated by a thermostatically controlled resistor. Measurements of the sample temperature with a calibrated copper–constantan thermocouple next to the cuvette show that temperatures are maintained to better than 0.5°C by the thermostat. This system is simple, reliable and relatively inexpensive, but not available commercially. The techniques of cryoenzymology and spectroscopy below 77 K are outside the scope of this chapter but further information can be found in ref. 11.

3.4.3 *Low-temperature Fluorescence*

Generally fluorescence intensity increases as the temperature of the sample is reduced. This effect arises because quenching processes (see Chapter 1) are usually much more temperature sensitive than is fluorescence emission. Indeed cooling to liquid nitrogen temperatures may increase fluorescence emission by an order of magnitude, an effect which can be useful if measurements need to be recorded from a weakly fluorescent material, for example the flavoproteins of the mitochondrial respiratory chain. It is unusual for commercial instruments to provide low-temperature accessories but there seems no reason why the system described above should not be used in the 'front face' mode, where fluorescence is excited and detected at the surface of the sample cuvette.

3.5 **Cuvettes**

3.5.1 *Pattern*

The most common type of absorption (and fluorescence) cell or cuvette is the square cell, open at the top, with a path length of 10 mm and working volume of 2–3 ml, fluorescence cuvettes being polished on all faces. Manufacturers' catalogues illustrate a mouth-watering collection of designs for special purposes, with path lengths ranging from 1 to 100 mm. The essential feature is parallelism of the two end plates which limit the path lengths. The magnitude of the other horizontal axis (i.e. perpendicular to the light beam) is dictated by the volume of the sample and the desirability of keeping the light beam clear of the side walls, where intense reflection can occur. In 'semi-micro' cuvettes (i.e. path length 10 mm, capacity ~0.65 ml), the side walls can be perilously close to the light beam and should be blackened or 'masked'. It is a useful exercise to determine the minimum volume that can be used with each set of cuvettes; this is done by gradually adding a coloured or turbid sample until the absorption (or fluorescence) reading is constant. The working volume can sometimes be reduced further by including a spacer below the cuvettes.

3.5.2 *Mixing*

Cuvette lids or, better, ground stoppers, prevent evaporation of solvent and/or contamination of the cuvette contents. Like cuvette stands, they are useful but surprisingly under-used. Tapered Teflon stoppers can be drilled to accommodate the fine needle of a syringe such as those supplied by Hamilton. Mixing of cuvette contents can be problematic, especially when the cuvette is almost full and a lid is fitted; a few glass beads (small enough to lie below the light path) can provide sufficient agitation when a cuvette is inverted. An alternative is to use the mixers or 'plumpers' available, for example, from Calbiochem or Clinicon International GmbH. They can be loaded with up to about 100 μl of an addition, then rotated in the cuvette between thumb and forefinger to provide simultaneous and fairly rapid addition and mixing; for semi-micro cuvettes, the 'platform' of the mixer can be trimmed with a blade.

Where the contents of a cuvette must be continuously stirred during the scanning of a spectrum, stirring blocks about 1 cm square and 0.5 cm high are available to fit below 1 cm cuvettes (Temtron Electronics Ltd., Centronic Sales, New Addington, Croydon, UK) and drive a tiny magnetic bar within the cuvette. The blocks are energised by a bulky power supply outside the spectrophotometer. Although suitable for modest stirring of small volumes, they are inadequate for mixing the contents of a larger cell such as that required in potentiometric redox titrations (see Section 3.5.4 below). Here, a magnetic stirring bar (about 1 cm long) rotates around a horizontal axis just above the light beam (13). The bar can be driven by a button magnet mounted on the shaft of a small 12 V variable-speed electric motor.

3.5.3 *Optical Material*

The choice of glass, silica (quartz) or plastic cuvettes will be primarily dictated by whether they will be used in the u.v. or visible spectral regions. Only silica is suitable below about 360 nm. If funds permit, such cuvettes circumvent the frustration of selecting an unsuitable cuvette for, say, scanning a haemoprotein from its variable absorption bands into the u.v. for determination of a haem (410 nm)/protein (280 nm) ratio. Plastic disposable cuvettes are cheap, popular and can give acceptable results. Their less-than-perfect optical faces may be of little consequence when working with highly light-scattering samples. However, the variation of absorbance reading found when plastic cuvettes are repeatedly replaced in the cuvette holder (usually to determine the end-point of a reaction) can be unacceptably large. The problem is most acute when the plastic cuvettes only loosely fit holders designed for glass or silica cuvettes which have slightly the greater dimensions.

3.5.4 *Special Applications*

Examples of cuvettes for special purposes and which must usually be constructed by the user are those for retaining standard e.p.r. tubes in a single beam, for low-temperature work and photodissociation spectra and for potentiometric titrations (see ref. 11).

3.5.5 *Cleaning*

The importance of clean cuvettes is self-evident, particularly in fluorescence applications where exogenous quenchers may severely affect the results obtained. Routinely, cuvettes should be emptied immediately after use, rinsed repeatedly in the solvent (e.g. water), then with clean ethanol or acetone and dried with low pressure air or nitrogen from a cylinder. It is prudent to install a filter (such as those with pore sizes of 0.45 μm used in bacteriology) in the gas line. Cotton wool 'buds' can also be useful for dislodging interior, stubborn marks and for drying. The outside optical surfaces should be polished with clean lens tissue. Note that plastic 'squeezy' bottles generally used for solvents contain plasticisers such as butyl phthalate, which can interfere with critical u.v. spectra and provide an unrecognised source of fluorescent material.

The Perspex windows of low-temperature cuvettes easily become scratched and will eventually crack. However, provided they are still liquid-tight (at least for as long as it takes to plunge them and their contents into liquid nitrogen) they are quite serviceable in this condition. The opacity of the Perspex is negligible compared with that of the frozen sample.

Neglected cuvettes that cannot be dismantled for cleaning may require soaking overnight in concentrated sulphuric acid containing a few crystals of dichromate or permanganate, or boiling in distilled water containing a laboratory detergent such as Teepol or Decon. Fluorescence cuvettes cleaned in acid dichromate or permanganate should be subsequently soaked in neutral solutions of EDTA to remove traces of heavy metals which are potent quenchers of fluorescence.

3.6 A Detailed Example: Recording of a Cytochrome Difference Spectrum (Reduced Minus Oxidised)

The most common method of cytochrome spectral analysis is the recording, with a split-beam spectrophotometer, of a spectrum that represents the difference between an oxidised and a reduced sample. Such spectra may provide data on the identity and the amount of the cytochrome types present, and may be obtained with suspensions of intact cells, subcellular fractions derived therefrom or purified preparations.

3.6.1 *Wavelength Range*

The most useful absorbance bands of reduced cytochromes (called γ, β, α, from low to high wavelengths, respectively; *Figure 1*) lie between 400 and about 650 nm. In this visible region, the tungsten source is appropriate and few constraints are placed on the light-transmitting material of the cuvettes, so that low temperature and other special purpose cells can be used.

3.6.2 *Optical Configuration*

Such a difference spectrum is generally recorded in the split beam mode with an oxidised sample in the reference position and a reduced sample in the 'test' position. If a dual-wavelength instrument is used, an absolute spectrum of the reduced form may be adequate or else a difference spectrum must be obtained by recording separately the spectra of the oxidised and reduced forms in the single sample position and then computing the difference.

3.6.3 *Instrument Settings*

A compromise must be reached in the choice of scan speed, slit width and response time, these factors being inter-related (*Figure 11*). 'Typical' or starting values for these parameters might be 1 − 4 nm/sec, 2 nm spectral band width and 0.1 sec, respectively.

3.6.4 *Preparation of Samples*

Full reduction of a sample can generally be achieved by adding a few grains of $Na_2S_2O_4$ to a cell or membrane suspension, but may take several minutes for completion. It is important that the dithionite should be fresh. It is useful to split a newly purchased

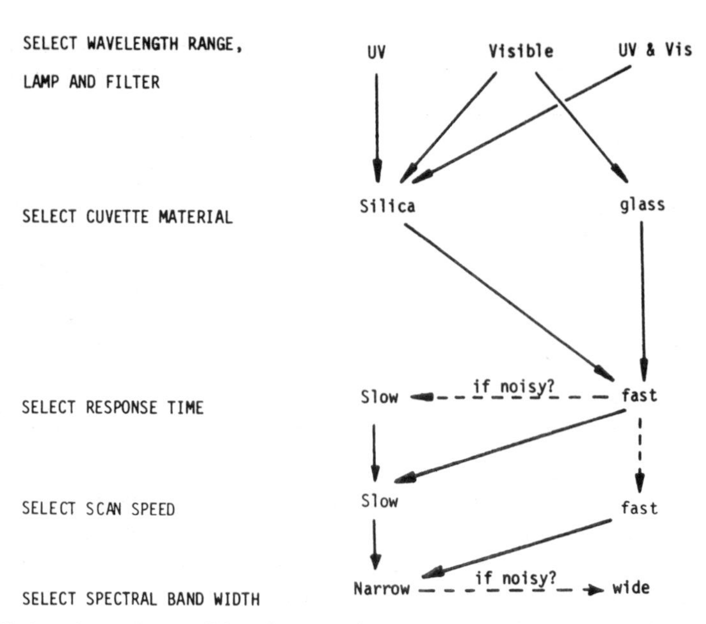

Figure 11. Choice of operating conditions for spectral scanning. Solid lines indicate preferred routes.

sample into small aliquots in, say 5 ml screw-capped bottles and to use each for only a month before discarding. Alternatively, the preparation can be allowed to become anoxic by respiration of a suitable oxidisable substrate. Sodium borohydride is also a useful reductant.

Oxidation may be achieved by using an exogenous oxidant or by vigorous aeration of the sample just prior to scanning the spectrum. Difficulty can be experienced if significant levels of endogenous reductants are present or if a respiratory chain is terminated by an oxidase with a particularly high O_2 affinity or is capable of particularly rapid rates of electron flux. In such cases, O_2 may be generated in the sample by adding H_2O_2 and catalase, although preparations of cells or sub-cellular fractions may exhibit sufficient endogenous catalase activity. Alternatively, potassium ferricyanide (ferro/ferricyanide +0.43 V), hexachloroiridate ($IrCl_6^{2-}/IrCl_6^{2-}$, +1.09 V), or ammonium persulphate ($S_2O_8^{2-}/2SO_4^{2-}$, +2.0 V) may be added as oxidants E°′, values in parentheses). The first two are highly coloured, however: aqueous solutions of ferricyanide absorb strongly between 370 and 460 nm (λ_{max} 420 nm), whilst hexachloroiridate solutions have complex spectra in this region with λ_{max} at 488 nm. Ammonium persulphate is colourless. Addition of these oxidants to final concentrations of about 2 mM is usually satisfactory, although it is common practice to add 'a few grains' to 1–4 ml samples.

Be wary of the problems of incomplete cytochrome reduction or oxidation by these reagents in quantitative work and of the danger of inadequate buffering against the pH changes that they may cause. The latter problem is especially acute in the case of dithionite, whose oxidation products are highly acidic. In addition, certain oxidases form spectrally distinct complexes with oxygen and peroxides, so that addition of these reagents does not yield the fully oxidised species. For example, cytochrome oxidase

Table 1. Summary of Factors that may Affect the Positions of Absorption and Fluorescence Maxima.

Factor	*Wavelength shift*
1. Instrument miscalibrated	Either direction
2. Scan speed too fast 3. Instrument response too slow	In direction of scanning, e.g. red shift in the case of scanning from low to high wavelengths
4. Temperature	Generally blue as temperature decreases, by 1–4 nm, depending on extent of change
5. Extreme slope of baseline	In direction of slope (e.g. red shift when baseline sloped down at high wavelengths)
6. Chemical environment of chromophore	Varied; see text for examples

d in *Escherichia coli* (and other bacteria) readily reacts with O_2 to give a stable, oxygenated form with a distinctive absorption band at about 650 nm. Thus, reduced *minus* oxidised (by aeration) difference spectra show a trough at this wavelength, whereas ferricyanide-oxidised samples are almost featureless in this region. Further details of these procedures and references are given by Jones and Poole (11).

4. FACTORS AFFECTING THE POSITION OF ABSORPTION AND FLUORESCENCE MAXIMA

Fundamental to the use of spectra for quantifying and identifying chromophores is an appreciation of factors that can alter the expected position of absorption or fluorescence maxima. *Table 1* summarises the operating conditions that can shift maxima to higher and lower wavelengths.

The fluorescence of many species is a sensitive indicator of the chemical environment of the fluorophore and changes in the environment may lead to substantial changes in fluorescence emission. This property has been used to explore the microenvironments within proteins and, particularly, biological membranes (14). Ligands that bind to the chromophore may also affect its absorption characteristics and this has been exploited extensively in studies of haemoproteins from which the following examples are drawn. The haem iron of cytochromes, haemoglobin, oxygenases and hydroperoxidases frequently binds ligands resulting in a characteristic absorbance change.

4.1 Carbon monoxide

CO binds to the reduced form of cytochrome oxidases, certain other cytochromes and haemoproteins, acting as a competitive inhibitor with respect to oxygen and forming a reduced cytochrome–CO compound which may be photodissociable (for a recent review, see 15).

The most common practical application of these important properties is the recording of reduced-plus-CO *minus* reduced difference spectra ('CO difference spectra'; *Figure 3*). Here, the difference in absorbance between two reduced samples of the preparation is recorded using a split beam or dual-wavelength scanning spectrophotometer to give a baseline. This is not always a trivial matter and is sometimes facilitated by prior reduction of a single sample with substrate or dithionite before the sample is carefully split between two matched, dry, clean cuvettes, taking care to avoid aeration during transfer. The 'front' or 'sample' suspension (in a split beam experiment) is bubbled with a fine stream of CO for a minute or so. If the top of the cuvette is smeared or sprayed with silicone anti-foam, problems with excessive bubbling and frothing of concentrated protein are circumvented. The spectrum representing the difference between the samples is again recorded and, after baseline correction, if necessary, may reveal a complex spectrum comprising peaks and troughs. For example, in the classical case of cytochrome a_3, the CO-binding component of cytochrome oxidase in mitochondria and certain bacteria, a peak in the difference spectrum at about 430 nm results from the characteristic absorbance maximum of the CO complex (in the 'front' cuvette) whilst a trough at 445 nm arises from the corresponding loss of absorbance of the reduced form in this cuvette, i.e. the peak absorbance of the CO-reactive form prior to bubbling with CO. Exploitation of the light-reversibility of CO-binding in photodissociation spectra is described below.

4.2 **Cyanide**

The reactions of cyanide with cytochromes are more complex than those of CO and less well defined, even in the case of the mitochondrial oxidase. Here, cyanide reacts with both oxidised and reduced forms; cyanide reactivity is most simply investigated in reduced-plus-cyanide *minus* reduced difference spectra analogous to those described for CO above.

4.3 **Photodissociation Spectra**

A photodissociation spectrum generally takes the form of an inverted CO difference spectrum, being obtained by computing the difference between the CO-liganded form of the sample (a haemoprotein) and the sample after photodissociation of the bound ligand. The rates of recombination of the dissociated ligand with the haem are such that these spectra are generally recorded at sub-zero temperatures, where the ligand re-binding is slowed and the difference spectrum can be obtained with the response time of a conventional split beam or dual-wavelength scanning instrument. The temperature required varies between haemoproteins, such that 173 K is adequate for many cytochrome oxidases, but temperatures near liquid helium (4 K) are needed for haemoglobins. The apparatus and protocol developed by Chance (12) for cytochrome oxidases is outside the scope of this chapter (but see 11) as are the cryostats needed for liquid helium work. The CO-binding characteristics of many cytochromes and cytochrome oxidases, however, are such that photodissociation spectra can be obtained using commercial 77 K accessories. Typically, one might record at 77 K the spectrum of the CO-liganded, reduced sample, with an oxidised, reduced (or second CO-liganded) sample in the reference position and then remove the cuvette holder to a second unsilvered or partly silvered Dewar for photolysis with the focussed beam of a 200 W projector or a photographic flash gun. If the reference cuvette also contains a CO-liganded sample, this, of course, must be masked during photolysis. The post-photolysis spectrum is run and from it is subtracted the pre-photolysis spectrum to obtain the photodissociation spectrum.

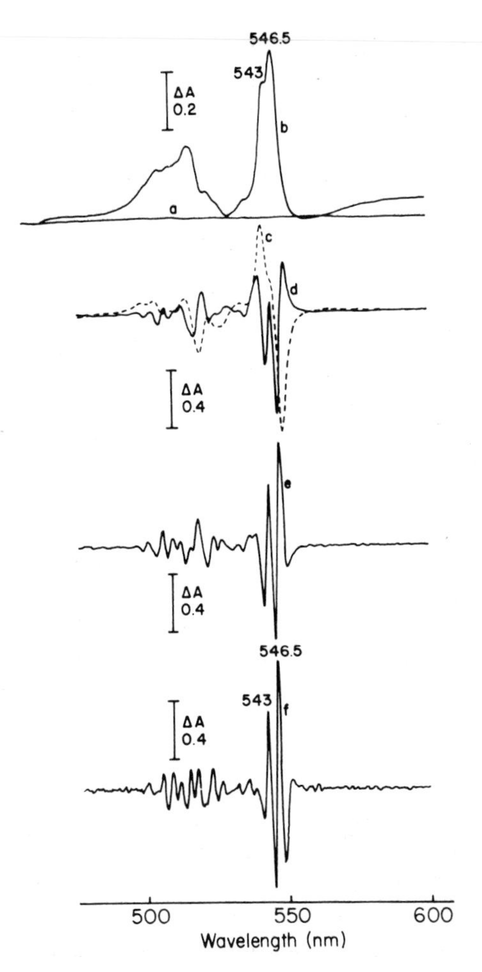

Figure 12. Reduced *minus* oxidised difference spectrum of cytochrome *c* and its first, second, third and fourth derivatives. The sample was the same as in previous figures. The spectra were recorded in the split beam mode at 77 K in a Hitachi 557 spectrophotometer. The baseline (**a**) was obtained using the baseline correction facilities with an oxidised sample in each cuvette. The difference spectrum was obtained by replacing the front cuvette with one containing a reduced sample (**b**). The derivative spectra [first through to fourth: (**c**) to (**f**), respectively] were plotted directly by the instrument during subsequent scans. The scan rate was 0.5 nm/sec, the spectral band width 0.5 nm and the response time 'fast'. The differentiating interval was approximately 5 nm throughout.

5. DERIVATIVE SPECTRA

In its simplest form, a derivative spectrum is a plot of $dA/d\lambda$ versus λ, rather than A versus λ. In the derivative spectrum of a single, symmetrical absorbance peak, the trace crosses the 'baseline' near the position of the original absorbance maximum. More complex forms arise from two or more bands close together; the general appearance and interpretation of such spectra will be familiar to e.p.r. spectroscopists, a technique where, for instrumental reasons, the first derivative is routinely plotted. When the first derivative spectrum is subjected to derivatisation, we obtain the second derivative and, by similar

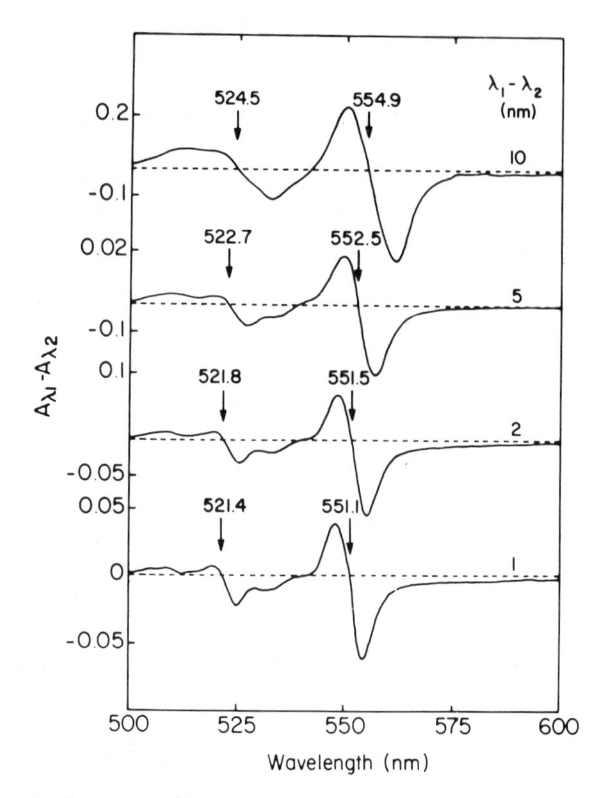

Figure 13. First derivative spectra of ferrocytochrome *c*. Spectra were recorded directly with an Applied Photophysics dual wavelength spectrometer with both monochromators set to scan. The wavelength offset of the monochromators ($\lambda_1 - \lambda_2$) and the position of the apparent absorbance maxima (the points at which the trace crosses the baseline) are indicated in each spectrum.

procedures, the third and fourth. The first and third derivatives have the general characteristic described above, i.e. crossing of the 'baseline' at the original A_{max}. The second and fourth derivatives retain the band-like appearance of the original spectrum; in the second derivative the peak is inverted to give a trough at a position close to the λ_{max} of the original absorbance band, whilst in the fourth derivative the signal reverts to a peak at this position, and for this reason the fourth derivative is sometimes considered the most useful. In the second through to fourth derivatives, 'wings' or small inflections appear on each side of the main signal, complicating the spectrum's form (and interpretation).

The main application of higher derivative spectra is in the resolution of partly overlapping absorbance bands, a feature made possible by the band-narrowing that occurs on successive derivatisation. This can be seen in the derivative spectra of pure cytochrome *c* shown in *Figure 12*. In the 'native' (reduced minus oxidised) spectrum, the split of the α-band is barely visible (and indeed would be lost with less favourable scan rates or spectral band width). The apparent band widths of the two components of this band decrease on successive derivatisation, reaching optimal resolution in the fourth derivative.

The problems of this type of analysis are legion and are discussed more fully elsewhere (16–18). Chief among them are:

(i) slight shifts of the band positions in the higher derivative spectra from their true positions;
(ii) the generation of spurious derivative bands by two appropriately spaced bands;
(iii) the suppression of peak heights in the derivative spectra by overlap with 'wings' or troughs arising from adjacent bands;
(iv) the marked discrimination in favour (i.e. enhancement) of narrow bands;
(v) limitations of detectability of the highest derivative bands by noise in the original data (18).

Despite these problems, the methods can yield invaluable information when the user is fully aware of the potential artefacts and the resolution limits. Examples of applications to the cytochromes of bacteria are given in ref. 11.

There are two main ways of computing higher derivative spectra. First, some instruments allow the direct plotting of the first and sometimes higher order derivatives. In the Hitachi 557, the integral microcomputer facilities can calculate and plot first through to fourth derivatives with no optical or electrical change of the operating procedure. The user selects the order of derivatisation required and the differentiating interval, the interval over which $dA/d\lambda$ is calculated. The spectra in *Figure 12* were recorded with this instrument, using a differentiating interval of approximately 4 nm. When the differentiating interval is too small, the derivative spectrum is suppressed. The derivative bands intensify when the differentiation is performed over longer intervals but there is a progressive shift in the positions of the derivative peaks (not shown). Conditions appropriate to specific needs must be determined empirically. First derivative spectra only can be obtained with dual-wavelength devices by scanning the sample with both monochromators but set slightly out of phase, typically by 1–3 nm. The phase displacement of the monochromators affects the spectrum in the same way as direct selection of the differencing interval (*Figure 13*). It is important to note that in this dual-wavelength mode, the derivative spectrum is of *one* cuvette, whereas the Hitachi allows plotting of the derivatives of a *difference* spectrum.

The second general approach, which can be implemented with a less costly spectrophotometer, is to digitise the data and transfer it to a microcomputer or to paper tape for subsequent computation. The early work of one of us involved interrogation of the analogue output from the spectrophotometer by a digital voltmeter and data logger unit which transferred absorbance values to a paper tape punch; the paper tape was analysed by a mainframe computer and the numerical analysis plotted on an X–Y recorder under computer control. On-line data sampling and computation are now more widely available. The algorithm used was that of Butler and Hopkins (16,17), one important feature of which is the appreciable improvement in signal/noise ratio when the four differencing intervals used in computing the fourth-order finite difference spectrum are similar but not equal. This facility is not available in commercial instruments such as the Hitachi 557. Manipulating the spectra on a micro- or mini-computer also allows the summing of several (n) spectra (in which case the signal/noise ratio will be improved by a factor of $\sqrt{n}$) and the application of smoothing routines. The noise level, particularly in the higher derivative spectra, often limits their usefulness and such measures, as well as the proper choice of operating variables, especially of slow scan speeds, needs serious consideration.

6. ACKNOWLEDGEMENTS

R.K.P. wishes to thank Dr C.A.Appleby for expert advice on many facets of spectrophotometry and for access to the Hitachi Perkin-Elmer 557. The Royal Society and SERC are thanked for funding the application and development of some of the techniques described.

7. REFERENCES

1. Bell,J.E. and Hall,C. (1981) In *Spectroscopy in Biochemistry*. Ball,J.E. (ed.), CRC Press Inc., Boca Raton, Florida, Vol. **1**, p. 3.
2. Brown,S.B. (1980) In *An Introduction to Spectroscopy for Biochemists*. Brown,S.B. (ed.), Academic Press, London, p. 14.
3. Edisbury,J.R. (1966) *Practical Hints on Absorption Spectrometry (Ultra-violet and Visible)*. Hilger and Watts, London.
4. Gratzer,W.S. and Beaver,G.A. (1984) In *Techniques in the Life Sciences, B1/1 Supplement, Protein and Enzyme Biochemistry BS109*. Elsevier Scientific Publishers Ltd., p. 1.
5. Wood,W.A. (1981) In *Manual of Methods for General Bacteriology*. Gerhardt,P. (ed.), American Society for Microbiology, Washington, p. 286.
6. Parker,C.A. (1968) *Photoluminescence of Solutions*. Elsevier, Amsterdam.
7. Miller,J.N. (1981) *Standards in Fluorescence Spectrometry*. Chapman and Hall, London.
8. Burgess,C. and Knowles,A. (1981) *Standards in Absorption Spectrometry*. Chapman and Hall, London.
9. Morton,R.A. (1975) *Biochemical Spectroscopy*. Adam Hilger, London, 2 volumes.
10. Wilson,D.F. (1967) *Arch. Biochem. Biophys.*, **121**, 757.
11. Jones,C.W. and Poole,R.K. (1985) In *Methods in Microbiology*. Gottschalk,G. (ed.), Academic Press, London, Vol. **18**, 285.
12. Poole,R.K. and Chance,B. (1981) *J. Gen. Microbiol.*, **126**, 277.
13. Dutton,P.L. (1978) In *Methods in Enzymology*. Fleischer,S. and Packer,L. (eds), Academic Press, New York, Vol. **54**, p. 411.
14. Radda,G.K. (1971) In *Current Topics in Bioenergetics*. Sanadi,D.R. (ed.), Academic Press, New York, Vol. **4**, p. 81.
15. Wood,P.M. (1984) *Biochim. Biophys. Acta*, **768**, 293.
16. Butler,W.L. and Hopkins,D.W. (1970a) *Photochem. Photobiol.*, **12**, 439.
17. Butler,W.L. and Hopkins,D.W. (1970b) *Photochem. Photobiol.*, **12**, 451.
18. Butler,W.L. (1979) In *Methods in Enzymology*. Fleischer,S. and Packer,L. (eds), Academic Press, New York, Vol. **56**, p. 501.

CHAPTER 3

Spectrophotometric Assays

D.A. HARRIS

1. INTRODUCTION

1.1 Beer-Lambert Law

Probably the most common use of spectrophotometry in biochemistry is to measure concentrations in solution. Indeed, in so far as no biochemistry laboratory today is complete without a spectrophotometer, the spectrophotometric assay for concentration must be considered today as one of the corner-stones of biochemical practice. In an earlier volume in this series, for example, a series of spectrophotometric assays for enzymes, etc., was used to characterise subcellular fractions separated by ultracentrifugation (1). Such assays use the absorbance parameter A, where

$$A = \log \frac{\text{(incident light)}}{\text{(transmitted light)}} = \log \frac{I_0}{I} \qquad \text{Equation 1}$$

which is measured directly on modern spectrophotometers. The Beer-Lambert law states that, at reasonably low concentrations, A is proportional to concentration in solution (s)

$$A = \log \frac{I_0}{I} = ks \qquad \text{Equation 2}$$

The logarithmic relationship should be noted – when $A = 2$, for example, only 1% of the incident light is transmitted and normal laboratory instruments are likely to give inaccurate readings. As a rule of thumb, readings where A is greater than 1 are generally ignored in measurements of concentration, unless special precautions are taken (see Chapter 2).

1.2 To Separate or Not To Separate

Quantitative measurements may be approached in two ways. The first involves the refinement of separation techniques until all (significant) components of a mixture are separated and identified. Amounts can subsequently be measured by some non-specific method (conductance, u.v. absorbance, refractive index, etc.). A typical example in biochemistry would be the separation and quantitation of nucleotides (ATP, ADP, AMP, etc.) by h.p.l.c. using a u.v. absorbance detector (2).

This approach is to a large extent that of the analytical chemist, interested in measuring one component in the presence of a number of similar compounds. Quantitation by this approach rests on developing separation techniques, and will not be dealt with here.

The alternative is to use a specific detection method, and to detect or develop a measurable property of only one component of a complex mixture. In the ideal case, none of the other components of this mixture show or develop this property (nor, indeed, interfere with its development). Since spectrophotometry carries the ability to select a measuring wavelength (typically in the range of 200 – 900 nm), it is widely used in this approach where selectivity of measurement, as well as sensitivity, is required. The measured property, of course, is colour – or, strictly, absorbance. As an example, the measurement of bacteriochlorophyll concentration in a suspension of complex biological membranes from *Rhodospirillum rubrum* involves determination of absorbance at 880 nm (A_{880}), where no other components absorb (3).

Biochemical systems are most often suited to this second approach. First, biological systems tend to produce a finite range of well-defined chemical species, rather than the undefined mixture presented to the analytical chemist. Thus if we define an assay for glucose in blood serum, for example, we need hardly worry that other hexoses will be present to interfere. Secondly, we can exploit the specificity of enzymes to enhance the specificity of our measuring system. Firefly luciferase, for example, can readily detect traces of ATP in the presence of large amounts of ADP and GTP (4) that would 'swamp' an h.p.l.c. separation system.

1.3 **Light Scattering – Clarification of Turbid Samples**

This tacit rejection of the 'prior separation' approach does not rule out some preliminary treatment in certain spectrophotometric assays. Typically, this might involve clarification of the solution by centrifugation; to measure absorbance accurately it is necessary to eliminate light scattering, which also leads to a decrease in transmitted light and thus an artifactual, apparent increase in absorbance. In the measurement of chlorophyll from leaf chloroplasts, for example, the chlorophyll is extracted by 80% acetone and, before measurement, the suspension clarified by centrifugation (5) (compare above). In the measurement of serum enzymes, etc., erythrocytes are removed by centrifugation.

Specific examples of centrifugal 'purification' of proteins from interfering small molecules (or *vice versa*) are described below. However, techniques more complex than centrifugation at bench centrifuge speeds are avoided, to keep the assays as simple as possible.

2. GENERAL ASSAY DESIGN

2.1 **Theoretical Aspects**

The requirements of any assay are 5-fold – it should be *specific, sensitive, accurate, precise* and *convenient* (rapid and cheap). To some extent these requirements must be balanced. While obviously there is no point in having a simple assay which gives imprecise (i.e., widely scattered) results, it is less widely realised that there is equally little point in devising an exquisitely sensitive assay where multiple dilutions of sample, the need for special glassware, etc., make measurements complex and tedious. It should be emphasised that, in choosing one of the assays below, the *least sensitive* technique compatible with the experimental set-up should always be chosen.

Only the first two requirements will be explicitly dealt with in the experimental section here. Specificity will be discussed in terms of range of substances detected and, more particularly, what commonly found materials might give false positive results (i.e., induce a colour change themselves) or false negatives (i.e., prevent normal colour development). It is interesting that, in biochemical assays, specificity tends to be governed by factors prior to colour development (e.g., the specific enzymes employed) and the number of actual coloured products is relatively small. Reduced phosphomolybdate (blue) or reduced NAD^+ (yellow) are seen below to form a recurrent theme in biochemical assays.

Sensitivity will be given below in terms of g or mol quantities capable of detection by a standard laboratory spectrophotometer reading on a 0 – 1 absorbance unit scale. A range of techniques of differing sensitivities will be given where available. Detection limits on other instruments (fluorimeters, photon counters) are to a greater extent limited by instrument noise and are somewhat arbitrary – convenient values are given assuming no specialised (i.e. dedicated) apparatus is available.

2.1.1 *Accuracy and Precision*

Accuracy and precision are defined through a statistical analysis of the results obtained (6). A repeated assay on the same sample, for example, yields a range of values, for the parameters measured:

$$\bar{x} \pm \sigma \quad \text{(where } \sigma = \text{standard deviation)}$$

In a *precise* assay, σ is small compared with $\bar{x}$; typically $\sigma/\bar{x}$ should be less than 5%. In other words, the *scatter* of the results is small, and the method is *reproducible,* with good *resolution*. (Resolution is the ability to distinguish between two close values.) *Random errors* (pipetting, general laboratory technique, etc.) affect assay precision.

An *accurate* assay, conversely, is one where the measured value $\bar{x}$ corresponds closely to the true value X ($|X-\bar{x}|$ small). At a trivial (but not uncommon) level, we may have added insufficient colour reagent to react with all our substance to be measured and $\bar{x}$ will underestimate X (*Figure 1a*). More typically, colour development may be slow in samples of low concentration, but faster at high concentration. Thus by measuring at a fixed time, the low concentration values may be consistently underestimated (*Figure 1b*). Accuracy is affected by *systematic errors.* Accuracy is maximised by using linear relations for calibration (to aid extrapolation between measured calibration points). It is also maximised by measuring at or near the wavelength of maximal absorbance (λ_{max}), so that a small error in applied wavelength has a minimal effect on the value measured (*Figure 2*).

Rather surprisingly, accuracy is often less important in designing biochemical assays than precision. This is because we are concerned generally with relative rather than absolute quantities (or changes). In enzyme isolation, for example, a protein assay may be calibrated with albumin. The colour change observed for 1 mg albumin may be rather different from that observed with 1 mg of the enzyme under study (see below), but as long as the ratio of the changes is constant (i.e., the error is systematic, not random), this method can still be used to follow the purification (increasing specific activity) of this enzyme.

Accuracy, however, may be critical in some cases. In measuring stoichiometries of ligand binding to proteins, for example, it is essential to know the accurate ('true')

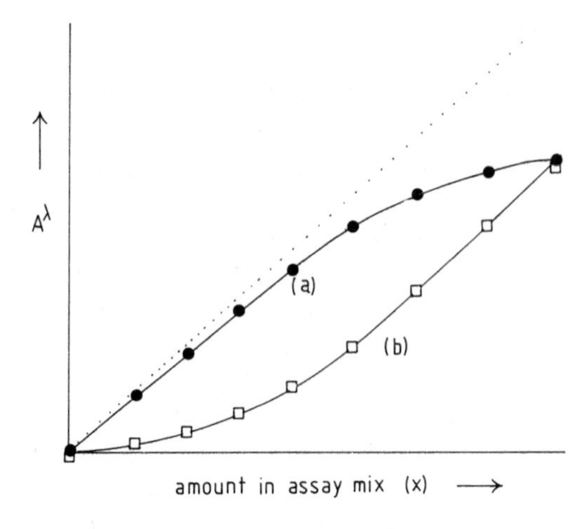

Figure 1. Non-linearity in colorimetric assays. Curve **(a)** (●-●) shows the response of the assay to increasing amounts of material (x) when insufficient colour reagent is present. The experimental line deviates from the optimal line (dotted) at high values of x, since all the colour reagent present has reacted. Curve **(b)** (□-□) shows the response when insufficient time is allowed for completion. The experimental line deviates most at low values of x, since the rate of reaction at low concentrations is lowest.

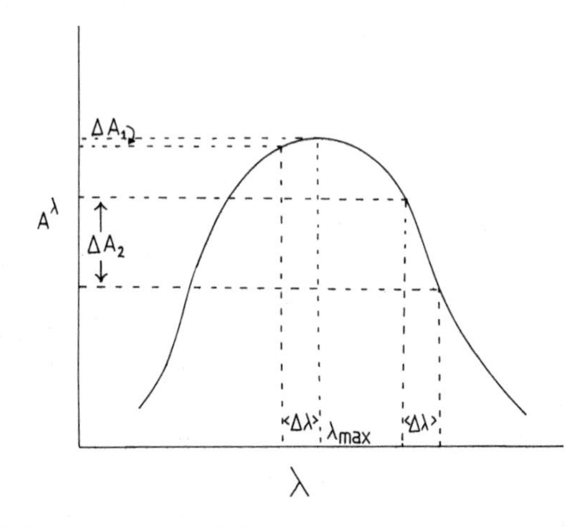

Figure 2. Selection of absorbance wavelength for measurement. A constant error in wavelength ($\Delta\lambda$) leads to a large error in absorbance (ΔA_2) when measurement is on the side of an absorbtion band, but only a small change if measurement is at the peak absorbance (ΔA_1).

molecular weight and concentration of the protein. In this case, careful and accurate calibration – preferably by some absolute method, using the pure protein under study – is necessary. The biochemical literature abounds with disputes over stoichiometries resulting from inaccurate (i.e., not absolute) calibration of protein assays. Methods of absolute calibration of macromolecule assays are suggested below.

2.2 Practical Aspects

2.2.1 *Volume*

An assay mixture contains, in principle, three components:

(i) sample (material to be assayed);
(ii) diluent;
(iii) colour-producing reagent (developer), in excess over the sample.

To aid comparison, the sum of the volumes of (i) and (ii) is kept constant (addition of more sample is balanced by adding less diluent). The volume of (iii) is always constant, leading to a constant total volume. In this case, the *absorbance* observed should be proportional to the *amount* of material added to the assay mixture – it is not necessary to calculate the concentration of this material within the assay mixture.

The total volume in the assays below is chosen to be 2.5 ml. This is convenient for most spectrophotometers/fluorimeters using 1 cm (path length) × 1 cm (width) cuvettes. Use of more material is wasteful; it will not fit in the measuring tube. The assays can be scaled down to a total volume of 1 ml for 1 cm × 4 mm (semi-micro) cells, by decreasing all three components of the assay mixture in proportion. Scaling down leads to a higher sensitivity with a minimal increase in complexity. However, using these smaller cuvettes, it must be ensured that:

(i) the light beam passes only through the liquid, not the cell walls (use a narrow slit width and/or black sided cells) (see Chapter 2);
(ii) cuvettes are well washed and dried between samples. The higher surface area/volume ratio in 1 cm × 4 mm cells makes cross-contamination more of a problem than in 1 cm × 1 cm cells.

2.2.2 *Diluent*

The diluent contains buffers, ions, etc., suitable to allow colour development. In the Biuret and Lowry protein assays for example (below), it consists of alkali (to keep the pH high), Cu^{2+} ions (to chelate to the protein) and tartrate (to keep the Cu^{2+} in solution at high pH). In the phosphomolybdate assay for phosphate, and the dye binding protein assay, the diluent contains strong acid, to keep the mixture at low pH. In enzyme-based assays it contains ions, buffers, etc., necessary for enzyme function.

To keep the assay conditions constant, besides a strongly buffered diluent, it is customary to keep sample volumes to 25% or less than the volume of diluent – much less in the case of a rate assay (see below). Even so, errors may arise – a sample containing, say, perchloric acid or EDTA may complex or precipitate essential components of the diluent, and checks should always be made to show such effects are unimportant in a system under study.

2.2.3 *Order of Addition*

Although order of mixing the reagents will rarely affect the final measurement, the colour-developing reagent, or an enzyme in an enzyme-based assay, is normally added last. This is because these reagents are likely to be least stable in the assay mixture, and their premature decomposition before addition of sample might lead to either abnormal colour development (high 'blanks') or slowed colour development. In some cases, the colour reagent may decompose so rapidly in the diluent that, if this reagent is added before the sample, *no* colour development will be observed. In these cases, rapid mixing is required during, or immediately after adding the colour reagent to the sample/diluent mixture (see Sections 3.4 and 3.6 below).

2.3 **Type of Assay**

Metabolite assays can be divided, in principle, into two types – 'end point' assays, where *all* the metabolite is converted into some coloured compound and 'rate' assays, where the metabolite is used to influence the *rate* of some measurable process. Typical examples might be the measurement of ATP concentration (i) by NADPH production from glucose using glucose-6-phosphate dehydrogenase (G6PdH, Section 7.2), and (ii) by measurement of light emission using firefly luciferase (see Section 8.2), respectively.

2.3.1 *End Point Assays*

End point assays should yield the most precise results, being unaffected by small changes in temperature, enzyme activity, time of measurement, etc., since measurement is made after reaction is complete (*Figure 3*, solid line). The requirements for the assays are given below.

(i) ΔG^0 must be large and negative, to ensure at least 98% reaction. Chemical trapping agents may be used.

(ii) The colour produced should be stable over the period of development and measurement. This may not be the case if the colours fade in time (protection from light often helps) or if one of the other components is unstable. In the estimation of ATP with G6PdH, as mentioned above, the final absorbance level drifts upwards because glucose itself is oxidised by G6PdH (*Figure 3*, dotted line).

(iii) Sufficient time is allowed for the end point to be reached. Surprisingly, this time is generally independent of initial concentration of the assayed component, providing this component is present in small amount relative to the other reagents. See Section 6.3 for an estimate of how long this time should be.

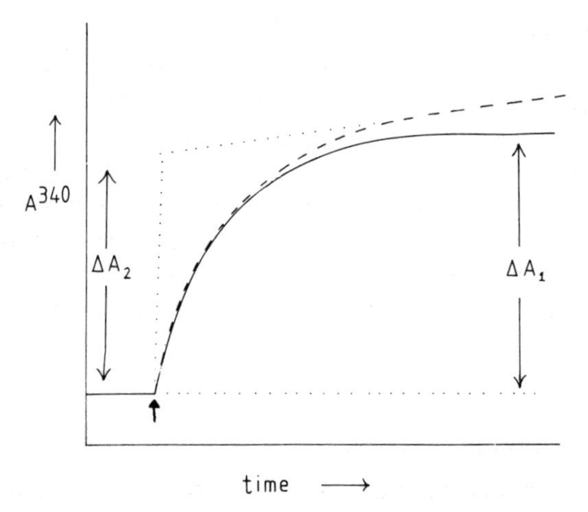

Figure 3. End point measurement in colorimetric assay. The solid curve represents an idealised reaction time course – the indicator enzyme, added at the arrow, leads to a steady change towards a fixed value, where ΔA_1 is proportional to the amount of assayed material. The dotted curve represents the situation where a slow secondary reaction is present, for example in the assay for ATP using glucose, hexokinase and G6PdH (Section 7.2). In this case, G6PdH will slowly oxidise unphosphorylated glucose, and thus NADPH production will continue slowly even after all the G6P has been consumed. This effect is minimised by keeping the glucose concentration low, and by extrapolating A to the time of initiation of the reaction (ΔA_2).

2.3.2 *Rate Assays*

Rate assays are less common, owing to their lower precision. Temperature, period of incubation and inclusion of interfering substances with the sample must be carefully controlled. However, rate assays are sometimes used when they lead to greater sensitivity – either because they allow use of a more sensitive measuring method (e.g., bioluminescent methods, Section 8.1), or because they allow 'amplification', that is, production of several mols of product per mol of compound assayed (e.g., assay of AMP using its indirect effect on glucose release from glycogen, ref. 7). Cycling methods are also useful, where the compound to be assayed is continuously regenerated at the expense of another easily measurable one, and this latter compound assayed (Section 10.1).

In all cases, it is important:

(i) that s_o, the concentration of the substance being assayed, is below the K_m of the measuring enzyme, so that the enzyme rate is approximately proportional to s_o;

(ii) that temperature, pH, salt concentration, etc., are strictly controlled, so that s_o is the only factor affecting the rate of the enzyme.

2.4 **Calibration**

It is customary to calibrate an assay by carrying out parallel estimations on solutions of known concentration. Conditions (volume, temperature, etc.) should be the same for the calibration as for the unknown samples. Calibration methods are suggested along with the assays described below.

3. SPECTROPHOTOMETRIC ASSAY FOR PROTEINS

3.1 **General Aspects**

At some stage in almost every biochemical investigation, it is necessary to measure the amount of protein present. This value may be used, for example, to calculate the specific activity of a purified enzyme, to calculate the amount of an enzyme to be added to a cell-free system, etc. Protein concentration could be estimated by dry weight measurements, but since this involves removal of all small molecules (buffers, ions) and water – a difficult task since most proteins are both charged and hygroscopic – this technique is seldom used, even for the calibration of reference solutions of proteins. Proteins are generally measured spectrophotometrically, using intrinsic (ultraviolet) chromophores such as aromatic residues (280 nm) or peptide bonds (215 nm) or by treatment with a reagent to yield a coloured or fluorescent product.

3.1.1 *Choice of Protein Assay Method*

Sensitivity is the overriding factor in choosing a protein assay, as most of the commonly used methods are reasonably convenient, and interfering substances can often be removed by acid precipitation of the protein prior to assay. Solubility of the protein may also be a relevant factor – in this case the Biuret or Lowry methods are generally used since they can be modified to include a protein solubilising step.

The assays are described in order of increasing sensitivity (2.5 ml volume) (see

Table 1. Detection Range of Commonly Used Protein Assays (cf. also ref. 9)

Method	*Range*
Biuret	1000 – 5000 μg
U.v. absorbtion	200 – 2000 μg
Lowry	5 – 40 μg
Dye binding	1 – 10 μg
Fluorescamine	0.1 – 2 μg

Table 1), and can all be modified to a total volume of 1 ml, giving an increase of 2.5 × in sensitivity. Clean tubes are obviously required for any assay, but while detergent-washed tubes will suffice for most purposes, for the measurement of 10 μg protein or less, acid-washed tubes (5 M HCl or 5 M HNO_3 overnight), pyrolysed tubes (400°C, 4 h) or new plastic tubes should be used, and care taken in handling to avoid sweat, saliva, etc., contacting surfaces in contact with assay solutions. Note that a fingerprint may contain up to 0.5 μg protein (8).

3.1.2 *Calibration of Protein Assays*

A stock solution of about 10 mg/ml bovine serum albumin in water is prepared by weighing, and can be stored frozen at this concentration or a 1:10 dilution. More dilute solutions lose protein, probably by adsorption, on storage. A more dilute calibration solution, if required, should be prepared daily from the stock solution.

The exact concentration of protein in this solution is determined spectrophotometrically (see comments on dry weight, above), using $A_{280}^{1\%} = 6.60$ (10). This will probably be 5 – 10% lower than the value calculated from the 'dry weight', due to hygroscopicity, etc. The concentration should be checked from time to time, to obviate effects due to surface denaturation of albumin, which occur if the solution is mixed too vigorously.

Calibration should be made with five or more different amounts of albumin over the range under study. After an initial calibration, only two standards need to be included in each subsequent batch of unknowns in spectrophotometric assays until the reagent batch is changed (or monthly, if convenient), since the calibration should be reproducible from day to day. Fluorimetric assays, however, need calibration each time they are used, being sensitive to environmental conditions and machine set-up.

3.1.3 *Absolute Calibration*

Since proteins vary in size and composition, a constant weight of different proteins may give different responses in any given assay. For example, an assay responding largely to aromatic groups will underestimate of the weight of proteins with a lower than average content of such groups. The calibration described above simply gives protein concentration in relative units, based on response relative to albumin. It cannot be too strongly emphasised that for absolute, stoichiometric measurements, some direct calibration is necessary using the protein of study (see above). A convenient method, if the molecular weight and amino acid composition of the protein is known, is to calibrate the protein assay relative to some spectrophotometric assay of a minor amino acid

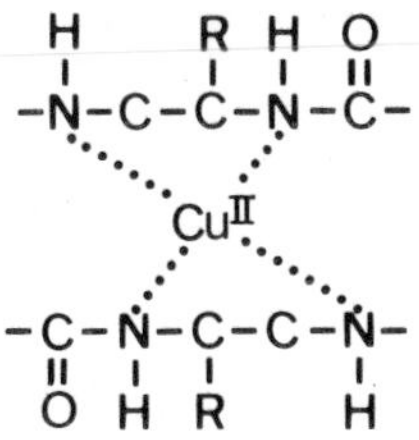

```
 H     R  H  O
 |     |  |  ||
-N--C--C--N--C-
   .       .
     Cu(II)
   .       .
-C--N--C--C--N-
 || |  |     |
 O  H  R     H
```

Figure 4. Tartrate complex of Cu^{II} in alkaline solution.

[cysteine (11), histidine (12), tryptophan (13)] in the pure protein. This said, however, the Biuret method, reacting as it does with peptide bonds, is perhaps the least susceptible of these methods to protein-protein variation and can be used almost as an absolute calibration in many cases.

3.2 **Biuret Assay** (14)

Cu^{2+} in alkaline solution complexes with nitrogen atoms in the peptide bonds of the protein. A purple complex is produced (*Figure 4*). Cu^{2+} is maintained in alkaline solution as the tartrate complex.

3.2.1 *Colour Reagent/Diluent*

(i) Dissolve 1.5 g of $CuSO_4.5H_2O$ and 6 g of NaK tartrate.$4H_2O$ (Rochelle salt) in 500 ml of boiled and cooled distilled water.

(ii) To this solution add (with mixing) 300 ml of CO_2-free NaOH (10%), followed by 200 ml of boiled and cooled distilled water.

This solution should be stable at room temperature for months. Decomposition of the reagent is indicated when an orange precipitate (of Cu^+) forms. The reagent must then be discarded, and fresh reagent prepared. If required, 1 g of potassium iodide per litre may be added to the reagent to prolong its shelf life.

3.2.2 *Method*

This method is for a colourless, soluble protein.

(i) Prepare a 0.2 ml sample (0.5 – 4 mg protein in 0.2 ml of water or buffer).

(ii) Add 2.3 ml of colour reagent/diluent.

(iii) Leave to stand at room temperature for 2 h. [For a faster result, heat to 80°C for 5 min, cool rapidly and stand at room temperature for 10 min.]

(iv) Read the A_{540} against a blank containing water (buffer) plus reagent.

(v) 1 mg of protein gives an A_{540} of about 0.1.

The colour is stable for several days if the solutions are kept dark. Interfering materials include ammonium ions at high concentrations (complex with Cu^{2+}) and reducing compounds (including reducing sugars) which reduce $Cu^{2+} \rightarrow Cu^+$ (orange precipitate).

3.2.3 *Modifications*

(a) *To remove soluble interfering materials*

(i) Make the protein solution 10% in trichloroacetic acid (TCA) by addition of a 50% w/v solution.
(ii) Place on ice for 5 min.
(iii) Pellet the precipitate by centrifugation and dry by aspiration.
(iv) Dissolve as in (c) below.

(N.B., if the protein concentration is less than 100 μg/ml, precipitation by TCA alone is incomplete. See Section 3.3.)

(b) *To remove coloured, protein-bound materials* (e.g., haem, chlorophyll, etc., ref. 15)

(i) Resuspend the precipitate from (a), using a vortex mixer, in 2 ml of 50% alcohol.
(ii) Add 1 drop of 50% TCA.
(iii) Keep the suspension at room temperature for 1 h and separate by centrifugation.
(iv) Treat the pellet similarly with 2 ml of 96% ethanol.

(c) *For insoluble proteins* [including pellets from (a) and (b) above]

(i) Add 0.2 ml of 0.25% deoxycholate in 0.1 M NaOH to the protein pellet.
(ii) Heat the mixture at 80°C for 5 min to dissolve the protein.
(iii) Cool the mixture and treat with colour reagent as above.

3.3 **Ultraviolet Absorption**

Aromatic residues absorb light in the near u.v. (280 nm) and peptide bonds in the far u.v. (215 nm). For a soluble, colourless protein, measurement of absorbance at either wavelength gives a measure of protein content.

3.3.1 *Method*

(i) If the protein solution is at all turbid, clarify it by centrifugation.
(ii) Measure the absorbance of protein in a buffered solution (pH 6–9) using the same buffer as a blank.

1 mg/ml protein gives $A_{280} \sim 1$ and $A_{215} \sim 15$.

The principle advantage of this method is that it is non-destructive, and it is frequently used for monitoring column eluates. Its disadvantage is its lack of specificity – many common buffers and biological materials (haem, nucleic acids, ATP, etc.) absorb strongly in the ultraviolet, especially at 215 nm. This method is useful for determining concentration if the protein is pure (especially if its aromatic content is known).

3.3.2 *Modifications*

To obviate the effects of other u.v. absorbing materials, measurement at two wavelengths is useful. For example, if nucleotides or a nucleic acid are present, protein content can be calculated using Equation 3 (16)

$$\text{protein (mg/ml)} \cong 1.55\, A_{280} - 0.76\, A_{260} \qquad \text{Equation 3}$$

For other absorbing materials, similar expressions can be derived if λ_{max} and ϵ_λ are known or independently measured.

3.4 'Lowry' method

By far the most quoted paper in the biochemical literature over the past 30 years has been 'Protein measurement with the Folin phenol reagent' (Lowry *et al.*, 1951) (17), reflecting the popularity of this type of assay. The method is two orders of magnitude more sensitive than the Biuret procedure, an advantage sufficient to outweigh its greater tendency to interference and to variability between proteins.

Like the Biuret procedure, the 'Lowry' method initially involves complexing the protein with Cu^{2+} in an alkaline solution. In addition, the copper appears to catalyse the reduction, by the tyrosine and tryptophan residues, of the phosphomolybdate/phosphotungstate anions in the Folin phenol reagent, added subsequently. This latter reaction leads to a blue colour (Section 12.1), which can be measured at 700–750 nm. Cu^{+} ions appear to be the catalyst in this reduction [see Section 3.4.3 (d) below].

3.4.1 *Reagents*

(a) 1% $CuSO_4.5H_2O$.
(b) 1% NaK tartrate.$4H_2O$.
(c) 2% Na_2CO_3 in 0.1 M NaOH.
(d) Commercial Folin-Ciocalteau reagent (sodium tungstate, sodium molybdate, phosphoric acid, hydrochloric acid).

On the day of use, mix solutions (a) and (b) in a 1:1 ratio and then dilute with 98 volumes of (c), to form the diluent. Dilute (d) 1:1 with distilled water on day of use, to form the colour reagent.

Solutions (a), (b) and (c) are stable at room temperature for months. Solution (d) is stable at 4°C for months.

3.4.2 *Method*

This method is for a colourless, soluble protein.

(i) Prepare a 0.2 ml sample (5–40 μg protein) in water (or buffer).
(ii) Add 2.1 ml of diluent (alkaline copper reagent).
(iii) Allow to stand for 10 min.
(iv) Add 0.2 ml of colour reagent and *mix immediately.*
(v) Allow to stand at room temperature (preferably in a dark cupboard) for 1 h and read the A_{750}, against a blank containing water (or buffer) and reagents. The colour is stable (in the dark) for several hours. 10 μg protein gives an A_{750} of about 0.1.

The absorption peak for the reaction is broad, and the linearity of the assay can be extended over the range 30–200 μg protein by measuring A_{550} rather than A_{750}.

Interfering materials include reducing agents (including thiols), phenols and many buffers (triethanolamine, Hepes, etc.) which lead to intense blue colorations in the blank), and other compounds (Tris, EDTA, sugars, glycerol) which slightly increase the blank but also decrease colour yield. If the concentration of the latter is small, the effects can be overcome by including the same compounds in the calibration; otherwise see below. Neutral detergents (e.g., Triton X-100) and large aliphatic bases (cyclohexylamine) cause precipitation of the reagent.

Removal of interference follows much the same lines as for the Biuret procedure.

3.4.3 *Modifications*

(a) *To remove soluble interfering materials* (18)

(i) Mix 1 ml of protein solution with 0.15 ml of 1% deoxycholate in 0.1 M NaOH.
(ii) Allow the solution to stand for 5 min.
(iii) Precipitate the protein/deoxycholate complex by addition of 0.2 ml of 50% w/v TCA, followed by 15 min on ice.
(iv) Pellet the protein by centrifugation, and remove the supernatant by aspiration.

Deoxycholate is necessary in this step as a carrier, since protein at very low concentrations is not precipitated by TCA. This method also allows protein to be concentrated into the assay medium, allowing measurement of protein concentrations as low as 5 μg/ml. To obtain a suitable amount of protein (5 μg) for assay using the standard sample volume of 0.2 ml, the original solution would have to be at least 25 μg/ml in protein.

(b) *To remove cloudiness caused by neutral detergents*

Inclusion of 3% dodecyl sulphate or deoxycholate in the alkaline copper reagent (diluent) will prevent neutral detergents causing cloudiness.

Note that attempts to precipitate protein using deoxycholate plus TCA, as in modification (a) above, is unreliable if neutral detergents are also present.

(c) *For insoluble proteins* [including pellets from (a) above]

Dissolve as for the Biuret procedure [Section 3.2.3 (c)] and then treat as for soluble protein.

(d) *Replacement of the Folin-Ciocalteau reagent* (19)

Cu^{+}, formed by reduction of Cu^{2+} in alkali by the protein, can be detected by complex formation with the specific Cu^{+} chelator, bicinchoninic acid (*Figure 5*). This can be contrasted with the reaction with Folin-Ciocalteau reagent, where Cu^{+} ions are used catalytically in the reduction of (pre-formed) phosphomolybdate.

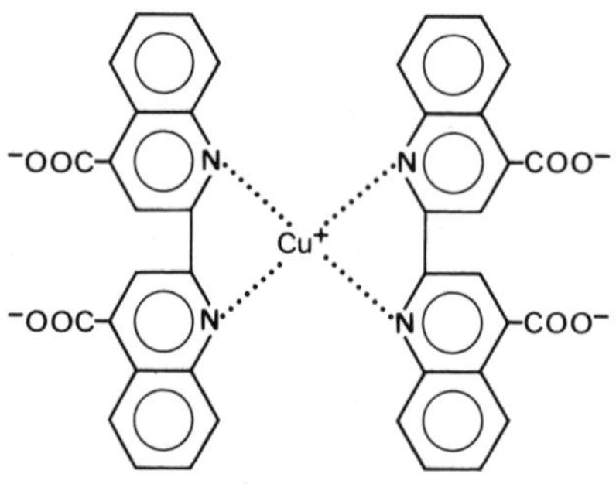

Figure 5. (Bicinchoninic acid)$_2$-Cu^{I} complex.

The following reagents are used:

(a) 4% $CuSO_4.5H_2O$;

(b) 1% bicinchoninic acid, 0.16% Na tartrate, 2% Na_2CO_3, 0.95% $NaHCO_3$ in 0.1 M NaOH (pH 11.25).

On the day of use, mix (a) and (b) in a 1:50 ratio. This forms the colour reagent/diluent solution.

Solutions (a) and (b) are stable for months at room temperature. Solution (b) can be obtained pre-mixed from Pierce Chemical Co. (BCA Protein assay reagent).

To carry out this modification, the following procedures should be performed:

(i) Prepare a 0.2 ml sample (5 – 40 μg protein).

(ii) Add 2.3 ml of colour reagent/diluent.

(iii) Allow to stand at room temperature for 2 h, and read the A_{562} (or, stand at 60°C for 30 min. High temperature incubations also increase sensitivity in this assay – only 1 – 5 μg protein need be used in this case).

(iv) The colour slowly increases with time, and all samples should be read within a 10 min period. 10 μg protein gives an $A_{562} \cong 0.1$.

Like the Lowry procedure above, this method is sensitive to interference by reducing agents, but it is much less sensitive to interference by neutral detergents or salts. It is also a simpler procedure, as the colour reagent is stable in the diluent solution. A drawback of this procedure is the dependence of A_{562} on temperature and time of incubation, which is not a feature of the other methods described here.

3.5 Dye-binding Assay

This assay is based on the colour change that occurs when Coomassie Brilliant Blue G250, in acid solution, binds to protein (20). The protonated form of Coomassie Blue dye is a pale orange-red colour. The dye binds strongly to proteins, interacting both hydrophobically and at positively charged groups on the protein (46). In the environment of these positively charged groups, protonation is suppressed and a blue colour observed (*Figure 6*).

3.5.1 *Reagents*

(a) Dissolve 100 mg of Serva Blue G in 50 ml of ethanol. Dilute the solution to

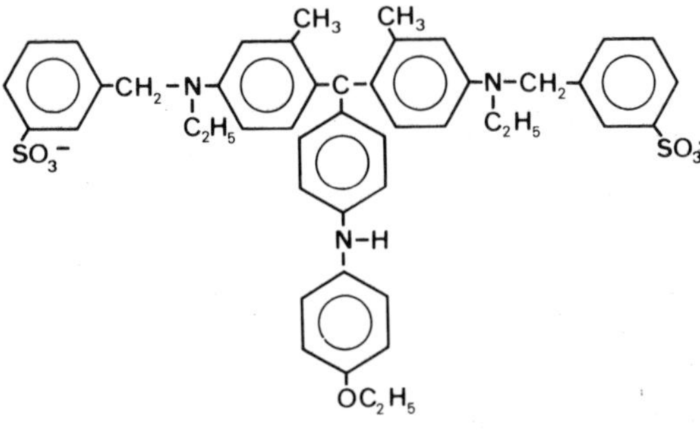

Figure 6. Unprotonated form of Coomassie Blue G250.

500 ml with water. [Serva Blue G is chosen because it is freely soluble in ethanol. Other dye preparations may require filtration before use.]

The concentration of dye in fresh stock solution (a) should be adjusted with 10% (v/v) ethanol to give $A_{550} \cong 1.18$ in the (a) + (b) mixture below. This ensures consistency between different batches of dye (21).

(b) 17% Phosphoric acid (commercial 85% phosphoric acid diluted 1:5). [0.3 M perchloric acid may replace 17% phosphoric acid, but is avoided here due to its oxidising properties.]

Solutions (a) and (b) are mixed in a 1:1 ratio on the day of use, forming the colour reagent/diluent. (Proprietory dye solutions are available from Biorad and Pierce Chemicals.)

Solutions (a) and (b) are stable for months at room temperature.

3.5.2 *Method*

This method is for soluble proteins only.

(i) Prepare a 0.2 ml sample (2 – 20 μg protein) in water (or buffer).
(ii) Add 2.3 ml of colour reagent/diluent (acid dye solution).
(iii) Mix and read the A_{595} immediately. The colour is stable for up to 30 min, after which precipitation of dye-protein complex may occur. 5 μg protein gives an A_{595} of $\cong$0.1.

Staining of cuvettes or reaction tubes due to adhering dye-protein complex may be removed after the measurements by washing with concentrated HCl or with methanol.

This assay is 2 – 3 times more sensitive than the Lowry assay (Section 3.4), and considerably faster and simpler. It is also less prone to interference by buffers, salts, reducing agents, etc. However, strong alkalis cause false positive results, and anionic detergents interfere.

3.5.3 *Modifications*

(a) *For insoluble proteins*

This assay is intolerant of anionic detergents (dodecyl sulphate, deoxycholate), and hence not very suitable for insoluble protein. However, the sample may be dissolved in Triton X-100 (0.2%), which does not interfere.

(b) *For very dilute solutions*

The volume of sample in the assay may be increased to 50% or even more of the assay volume. The stock dye solution must in this case be increased in concentration so that the final concentration (in the assay) is:

0.01% dye
0.8 M (5% v/v) ethanol
1.6 M (9% v/v) phosphoric acid (21)

Note that the sample should not be strongly buffered, or protonation of the dye will be suppressed and the blue colour will appear in the absence of protein.

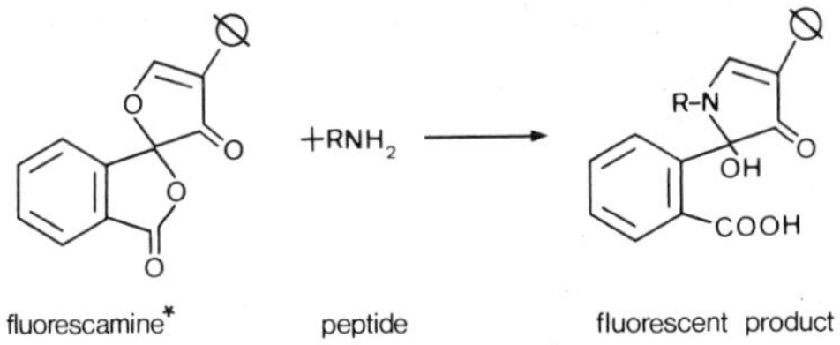

fluorescamine* +RNH$_2$ → fluorescent product
peptide

Figure 7. Fluorogenic reaction of fluorescamine with amines. Fluorescamine is available, under the trade name 'Fluram', from Roche Products Ltd., Diagnostics Department, 15 Manchester Square, London W1M 6AP.

3.6 **Fluorescamine Method (22)**

Fluorescamine, a heterocyclic dione, reacts with primary amines to form a fluorescent product (*Figure 7*). (Free NH_3 yields a non-fluorescent product.) The fluorescence of a solution containing protein plus fluorescamine is thus proportional to the quantity of free amine groups present. [Only this method, and A_{215} measurements can be used routinely for detection of small peptides (mol. wt. $\leq$10 000). The other methods described above give poor and variable results.] The reaction is carried out at pH 8.5, to suppress ionisation of the amine groups.

N.B. Fluorescamine hydrolyses quite rapidly in water to give non-fluorescent products ($t_{1/2}$ for reaction with peptides is 10 – 100 msec, hydrolysis takes 1 – 10 sec). **Keep water away from stock solutions of fluorescamine** – including wet pipette tips.

3.6.1 *Reagents*

(a) Borate buffer pH 8.5 (3 g of boric acid and 4.8 g of sodium tetraborate in 200 ml of water) (diluent).

(b) 7.5 mg of fluorescamine in 25 ml of acetone (AR) (colour reagent).

Solution (a) is stable for months at 4°C. Solution (b) is stable at room temperature, if kept free of moisture, but it is convenient to prepare it on the day of use, using a *dry* tube. AR acetone does not normally need to be specially dried.

3.6.2 *Method*

This method is for colourless, soluble proteins.

(i) Prepare a 0.2 ml sample (0.1 – 2 μg protein) in water (or buffer).

(ii) Add 2.1 ml of borate buffer.

(iii) Mix on a vortex mixer and, while still mixing, add 0.2 ml of fluorescamine solution.

(iv) Read the fluorescence after 2 min, using λ_{ex} = 390 nm, and λ_{emit} = 465 nm.

Because of the sensitivity of fluorescence to external influences (temperature, etc.), a full calibration should be made whenever this technique is used for assaying proteins. However, since the method is fast and simple, this is rarely a problem. It is convenient to adjust the fluorimeter sensitivity to read 100% for the maximal point on the calibration curve, and 0% for the blank (reagents + no protein).

The major source of interference is the presence of amine buffers (Tris, ammonium bicarbonate) in the protein solution. These must be avoided in preparing the sample. The importance of cleanliness in a highly sensitive assay has been emphasised above (Section 3.1).

4. ASSAYS FOR OTHER MACROMOLECULES

4.1 **Spectrophotometric Assays for Nucleic Acids**

DNA and RNA have a much narrower range of molecular properties than proteins. They can be extracted from cells with phenol, precipitated from solution with acid or ethanol, and re-dissolved in salt solutions prior to spectroscopic measurement, removing most interfering materials. The resulting nucleic acid can be measured either spectrophotometrically or fluorimetrically.

4.2 **Ultraviolet Absorption**

The nucleic acid bases absorb light strongly in the near ultraviolet (260 nm) and, in the absence of other absorbing material, can be determined by A_{260} measurement.

4.2.1 *Method* (See Section 3.3)

(i) To dissolve nucleic acids, 100 mM NaCl should be included in the buffer used.
(ii) 1 mg/ml of nucleic acid gives an A_{260} of $\cong$ 20. (Compare $A_{280} \cong 1$ for a protein solution of the same concentration.)

Interfering materials, such as protein, ATP, phenol and other aromatic molecules are usually removed in the preparation of the nucleic acid. For pure DNA, A_{260}/A_{280} = 1.8, and for pure RNA, A_{260}/A_{280} = 2.0. Values significantly less than this indicate that interfering materials are present.

4.3 **Fluorimetric Assay for DNA with Hoechst 33258**

The dye Hoechst 33258 [(2-[2-(4-hydroxyphenol)-6-benzimidazolyl-6(1-methyl-4-piperazyl) benzimidazole (23)] interacts with DNA with a many fold increase in fluorescence. The interaction is a physical one, and not yet well defined. AT-rich DNA gives a higher response than GC-rich DNA, and single-stranded DNA a lower response, and these factors should be taken into account when calibrating the assay.

4.3.1 *Reagents*

(a) 2 M NaCl containing 50 mM sodium phosphate (pH 7.4) and 2 mM EDTA (diluent – NaCl dissociates DNA-histone complexes).
(b) 200 μg/ml of Hoechst 33258 in water (colour reagent).

Solutions (a) and (b) are stable for months at 4°C. On day of use, dilute (b) 200-fold in solution (a).

A stock DNA solution (200 μg/ml) should be prepared in 100 mM NaCl for calibration. This solution can be stored frozen, but should not be repeatedly frozen and thawed. Store at −20°C in small aliquots.

4.3.2 *Method*

This method can be used directly to determine DNA concentrations in a tissue homogenate prepared by homogenisation of tissue in reagent (a).

(i) Prepare a 0.1 ml sample (0.1 – 1 μg DNA).
(ii) Add 2.4 ml of buffered dye dilution.
(iii) Incubate for 5 min at 20°C, and read the fluorescence at this temperature. λ_{ex} = 355 nm, λ_{emit} = 460 nm.

It is convenient to adjust the fluorimeter to read 100% with 1 μg of DNA present, and 0% with reagents only (see Section 3.6).

RNA does not interfere with this assay, since the response of the dye to RNA is only 0.25% that of DNA. Fluorescent materials, and turbidity in the sample, interfere with measurements. If the sample is itself fluorescent, a blank solution with sample plus buffer, but no dye, should be measured and subtracted from the actual reading. For turbid samples, see below.

4.3.3 *Modifications*

(a) *For turbid samples*

Tissue homogenates may be clarified by brief sonication (2 × 15 sec) or by centrifugation after homogenisation in the high salt buffer (a).

If the sample is still turbid, an internal standard may be included for calibration, viz. the assay mixture is read, a known amount (say, 200 ng) DNA added, and the sample read again.

(b) *For increased sensitivity*

The sensitivity of this method can be increased, with a sufficiently sensitive instrument, simply by using a 1:2000 dilution of dye solution (b) in the buffered salt solution (a) (i.e., a final dye concentration of 100 ng/ml). The sample, and calibration, should contain 10 – 100 ng of DNA.

4.4 Fluorimetric Assay for DNA Plus RNA with Ethidium Bromide (24)

The fluorescence of ethidium bromide is enhanced on binding either to DNA or to RNA. Enhancement probably involves intercalation of molecules of the dye between the nucleic acid bases.

4.4.1 *Reagents*

(a) Phosphate-buffered saline (PBS) (170 mM NaCl, 3.3 mM KCl, 10 mM Na_2HPO_4, 1.8 mM KH_2PO_4, pH 7.2).
(b) 100 μg/ml of heparin in PBS.
(c) 100 μg/ml of ethidium bromide in PBS.

Solution (b) is diluted 1:20 with solution (a) before use (diluent). Heparin is included in the assay to dissociate DNA-histone complexes.

Solution (a) is stored at 4°C as a 10-fold concentrate, for months. Solutions (b) and (c) can be stored at 4°C for up to 1 week.

A standard DNA solution is prepared and stored as above (Section 4.3).

4.4.2 *Method*

This method can be used directly to determine nucleic acid concentration in a tissue homogenate – see Section 4.3.

(i) Prepare a 0.2 ml sample (0.5–5 μg nucleic acid).
(ii) Add 2.2 ml of PBS/heparin diluent, and incubate at 30°C for 5 min.
(iii) Add 0.1 ml of ethidium bromide solution, and measure the fluorescence. λ_{ex} = 360 nm, λ_{emit} = 580 nm.

Calibrate as above (Section 4.3), except that 5 μg of DNA is used as the uppermost (100%) value.

4.4.3 *Modifications*

(a) *For turbid samples*

Samples should be clarified as far as possible by sonication and/or centrifugation. Internal standards (1 μg DNA) should be employed if the samples are still cloudy or coloured (see Section 4.3).

(b) *Determination of DNA or RNA*

(i) To one of duplicate samples in PBS/heparin diluent add 25 μg of (DNase-free) RNase.
(ii) Incubate the sample at 37°C for 30 min to destroy RNA.
(iii) Add ethidium bromide to both samples.
(iv) Measure the fluorescence.

Fluorescence of the digested sample gives the amount of DNA in the samples. The fluorescence difference between the two samples gives the amount of RNA in the sample. Note however that, weight for weight, double-stranded DNA enhances ethidium bromide fluorescence 2.17 times as much as RNA; thus if double-stranded DNA is used for the calibration, a correction must be applied to calculate RNA values.

5. SPECTROPHOTOMETRIC DETERMINATION OF COMPLEX CARBOHYDRATES

Complex carbohydrates are generally insoluble, and not directly suited to spectrophotometric assay. In general, they are first hydrolysed and the monosaccharides produced measured (see, e.g., Section 6, below). DNA and RNA can be similarly assayed, by measuring the carbohydrate produced on their hydrolysis (see ref. 1), but the techniques are long, complex, and less sensitive than the methods given above.

6. ASSAY OF SMALL BIOMOLECULES (METABOLITE ASSAYS)

6.1 Metabolite Assays versus Assay of Macromolecules

Assays of metabolites differ in several important aspects from the assay of macromolecules (above).

(i) The assays described above respond to all members of a class of macromolecules, e.g., all proteins, all nucleic acids, etc. When assaying metabolites – particularly if extracted from tissues – the assay must be specific for the metabolite under study. Such specificity is generally achieved by using an *enzyme-based assay*. (The macromolecular equivalent, where a test is for a specific protein, would most likely use an antibody-based assay.)

(ii) Above, the presence of small molecules, although possibly affecting the assay itself, does not affect the amount of macromolecule present. In tissue extracts, enzymes will be present whose purpose in life is to alter the amount of metabolite present. These must therefore be removed by deproteinisation – extraction in the presence of protein denaturants such as perchloric acid, TCA, $ZnCl_2$ (cold) or ethanol (hot or cold). Since these agents, by definition, denature enzymes, they must be removed as far as possible before enzymes used in the assay are added, e.g., by precipitation (neutralisation of $HClO_4$ by KOH yields insoluble $KClO_4$) (25), extraction into ether (TCA) (26) or evaporation (ethanol), or only very dilute extracts may be used for assay. Care is needed, however; some denaturants may alter the amount of metabolite present – acid may destroy labile phosphates, and will destroy NADH very rapidly (see Section 10.1).

(iii) The chemical reagents used above require strongly acid or alkaline conditions for the colour reaction. Enzymes, in general, function around pH 7 at room temperature, and any indicator/dye involved in metabolite assays must be stable/react around this pH.

6.2 Design of Assay

As examples, two enzyme-based assays for glucose will be described. [The reader is referred to a third – a purely chemical assay for reducing sugars in general (27).] The second, linked to reduction of a pyridine nucleotide, is an example of an important class of such linked assays and can be modified quite simply to detect other compounds. One advantage of an NAD(P)-linked assay is that its sensitivity can be varied over a wide range, depending on whether absorbance (detection range 0.1 – 1 μmol) or fluorescence (1 – 10 nmol) is measured.

The following principles of assay design will be maintained here, just as they were above.

(i) The total assay volume will be 2.5 ml (can be modified to 1 ml).

(ii) Sample size will be 0.05 – 0.5 ml. Smaller volumes lead to problems in dispensing, large volumes significantly affect buffering, ionic strength, etc., in the assay.

(iii) Assay pH will be between pH 6 and pH 9. Since both assays are end point methods, the enzymes involved need not be at their optimum pH, but they do need to be active.

(iv) If the other reactants are cheap and non-absorbent, they can be added in 10- to 100-fold excess, to (a) make ΔG more negative and (b) increase reaction rate.

(v) ΔG for the reaction should be such that greater than 99% reaction will occur, i.e., $\Delta G < -3$ kcal/mol. 'Trapping' reagents (e.g., semicarbazide for ketones) may be useful. Reactions driven by ATP (e.g., phosphorylation of an alcoholic group), and oxidations by O_2 normally are sufficiently exergonic.

(vi) The assay medium should contain not only enzymes and reactants, but any cofactors required by the enzymes, e.g., Mg^{2+} for kinases, K^+ for pyruvate kinase, and generally traces of EDTA to protect against heavy metal ions. Similarly, they should not contain enzyme inhibitors – enzymes packed as an ammonium sulphate precipitate may require dialysis, or centrifugation through a small Sephadex column (28) before use.

(vii) The assay is started by adding enzyme (or another labile component of the reaction mixture).

6.3 **Amount of Enzyme Required**

The amount of enzyme to be added is chosen so that the reaction is 'virtually' complete in a 'reasonable' time. If we take 'virtual' completion as ≥99% disappearance of substrate, and a 'reasonable' time for the assay as 5 – 10 min, we can calculate a nominal required quantity of enzyme.

For an enzyme reaction $-\frac{ds}{dt} = v = \frac{V_m . s}{K_m + s}$

As the reaction approaches completion $s \ll K_m$

$$-\frac{ds}{dt} = \frac{V_m}{K_m} s \qquad \text{Equation 4}$$

Integrating, to determine total time for a given change in s

$$-\int_{s_0}^{s_1} \frac{ds}{s} = \frac{V_m}{K_m} \int_0^t dt \text{ since } V_m, K_m \text{ are constants}$$

$$\ln \frac{s_0}{s_1} = \frac{V_m}{K_m} t \qquad \text{Equation 5}$$

For a decrease in s from 100% to 1%, in a time of 10 min (i.e., values we considered reasonable above)

$$2.3 \log \frac{100}{1} = 4.6 = \frac{V_m}{K_m} \times 10$$

or

$$\frac{V_m}{K_m} = 0.46 \quad (\text{min}^{-1}) \qquad \text{Equation 6}$$

For convenience, therefore, we chose an amount of enzyme (V_m = x units/ml assay medium) so that $\frac{V_m}{K_m}$ is between 1 and 10. So, for example, if K_m = 100 μM and our assay volume 2.5 ml, we would use about 1 U (μmol/min) enzyme, since

$$\frac{V_m}{K_m} = \frac{1 \ \mu\text{mol/min}}{2.5 \times 10^{-3}\ \text{l}} \times \frac{1}{100\ \mu\text{mol/l}} = 4(>0.46) \quad (\text{min}^{-1})$$

The volume of enzyme solution to be added can thus be calculated from the given specific activity of the preparation used. The following points are relevant to the above treatment.

(i) V_m refers to the activity of the enzyme in the assay medium. Conditions in the assay medium may not be identical to the conditions used by the manufacturer for his measurement of enzyme activity, and/or the enzyme activity may have decreased on storage. The activity of a stored enzyme, in the assay buffer, should be checked from time to time.

(ii) Note that few enzymes can be frozen and thawed in dilute solution without considerable loss of activity.

(iii) The reaction will, of course, proceed faster if more enzyme is added – rapidity must be balanced against expense.

(iv) Equation 6 contains no s_0 term, i.e., the amount of enzyme required is apparently independent of initial substrate concentration. This is surprising until it is realised that the calculation assumes $s_0 \ll K_m$. If $s_0 \gg K_m$, 99% reaction will be achieved faster than above, as more enzyme is working initially, and the time taken will indeed depend on s_0.

6.4 **Troubleshooting**

Always check the assay mixture first, using a suitable standard solution, to avoid wasting valuable sample. If an assay mixture fails to respond to the standard solution, the following steps should be carried out:

(i) Check the pH of the assay mixture in the cuvette (pH paper is generally sufficient).

(ii) Add in turn all reactants and co-factors to test if one has been inadvertently omitted. (Add to a single cuvette in the spectrophotometer.)

(iii) Independently check the activities of each enzyme used in the assay buffer. The commonest cause of an assay failing to work is denaturation of one of the enzymes involved, or omission (complexation – e.g., of Mg^{2+} with EDTA) of one of the enzyme's co-factors.

If the optical density of an assay solution drifts prior to starting the reaction carry out the following checks:

(a) Before the first assay

(i) Check the pH of the assay mixture. If it has dropped, bacterial contamination is probably the cause of drift.

(ii) Make up the assay mixture one component at a time in the cuvette, to check if one particular reagent is contaminated.

(b) In the second or subsequent assays

(iii) Wash the cuvette between assays with acetone or ethanol, and dry with a hairdryer, to eliminate carry over of enzyme activity between samples. Specific examples are given below in *Tables 2* and *3*.

7. DETERMINATION OF GLUCOSE

7.1 **Dye Oxidation Method**

Glucose is oxidised by glucose oxidase (a fungal enzyme) and O_2, and the H_2O_2 produced used by peroxidase to oxidise a leuco dye (2,2′-azino-di-(3 ethylbenzthiazoline) 6-sulphonate, ABTS) to a coloured (green) product. [ABTS is a trademark of the Boehringer Chemical Co. It replaces the (possibly carcinogenic) benzidine or dianisidine used previously.] This reaction is used semi quantitatively in commercial 'Glucostix'™, where the colour is produced on a paper strip.

$$C_6H_{12}O_6 + O_2 \xrightarrow[\text{glucose oxidase}]{} C_6H_{10}O_6 + H_2O_2$$

$$H_2O_2 + DH_2 \xrightarrow[\text{peroxidase}]{} 2H_2O + D$$ (DH_2, reduced dye; D, oxidised dye.)

7.1.1 *Reagents*

All reagents are dissolved in 0.1 M sodium phosphate, pH 7.0.

(a) 1 mM ABTS.
(b) 100 U/ml of glucose oxidase (GOD).
(c) 20 U/ml of peroxidase (POD).

These reagents can be stored at 4°C for up to 4 weeks.

7.1.2 *Method*

(i) Prepare a 0.2 ml sample (1–10 nmol glucose).
(ii) Add 2.1 ml of ABTS solution and 0.1 ml of peroxidase.
(iii) Incubate at 37°C for 5 min.
(iv) Add 0.1 ml of glucose oxidase.
(v) Incubate for 10 min and read the A_{420} against a blank containing reagents but no glucose.

The reaction is started not with peroxidase but with glucose oxidase, since the substrate of peroxidase, H_2O_2, is unstable.

Interfering materials include reducing agents (thiols, ascorbic acid), which are also oxidised by H_2O_2, or catalase (which may be a contaminant in peroxidase) which destroys H_2O_2.

7.1.3 *Modifications*

Glucose oxidase is very specific for glucose. Alternative oxidases (galactose oxidase, D-amino acid oxidase, etc.) enable other compounds to be measured. Any reaction producing H_2O_2 can also be monitored in this way.

7.2 **NADP-linked Assay**

Glucose is converted into glucose-6-phosphate (G6P) using hexokinase. G6P is then oxidised by $NADP^+$, using G6PdH, yielding NADPH. (These reactions constitute the first two reactions of the pentose phosphate pathway *in vivo*.) NADPH is measured by A_{340}.

$C_6H_{12}O_6$ + ATP → C_6H_{12}-O_6-Ⓟ + ADP (hexokinase)
$C_6H_{12}O_6$-Ⓟ + $NADP^+$ → $C_6H_{10}O_6$-Ⓟ + NADPH (G6PdH)

7.2.1 *Reagents*

(a) 200 mM Tris, 2 mM $MgCl_2$, 0.2 mM EDTA pH 8.0 (Buffer I) (phosphate buffer inhibits G6PdH).
(b) 200 mM MgATP (pH 7.0). Dissolve ATP (1 mmol) in 4 ml 250 mM $MgCl_2$, neutralise with 1 M KOH, and bring the final volume to 5 ml.

(N.B. The MgATP complex should be added already neutralised. If ATP is added to a buffer containing $MgCl_2$, $2H^+$/ATP is released as Mg complexes with the ATP, and the pH of the assay buffer falls.)

(c) 100 mM $NADP^+$ (pH 6.5) ($NADP^+$ should be kept in buffer below pH 7.0, or it will be destroyed).

(d) 20 U/ml of G6PdH.

(e) 100 U/ml of hexokinase.

7.2.2 *Method*

(i) Prepare reaction mixture containing 2.1 ml of buffer I, 0.05 ml of MgATP (200 mM), 0.025 ml of $NADP^+$ (100 mM), 0.1 ml of G6PdH and 0.2 ml of sample (40 – 400 nmol glucose).

(ii) Mix well, and incubate in the spectrophotometer (room temperature to 37°C) for 5 min. Measure the A_{340}.

(iii) Add 25 μl of hexokinase, and measure the A_{340} at intervals until steady.

(iv) Take the difference between the two values, ΔA_{340}.

$$\frac{\Delta A_{340}}{6.22} \times 2.5 = \mu\text{mol glucose present}$$

A blank should be prepared with all reagents but without glucose.

G6P does not interfere with the assay as described above, since it is completely reacted before the addition of hexokinase. However, if large amounts of G6P are present, measurement may be difficult against the high initial reading. (Note that any absorbance readings above 1.0 are likely to be inaccurate on a laboratory spectrophotometer.) Method 7.1 is more suitable if G6P is present in any amount. Other interfering materials are likely to be those which interfere with enzyme function – large amounts of EDTA, acid or perchlorate anions from the sample, for example.

Hexokinase has a fairly wide tolerance for hexoses, but G6PdH (and thus this method) is highly specific for glucose.

7.3 **Choice of Method**

Method 7.1 uses a dye, ABTS, which is both more stable and cheaper than $NADP^+$ and, in view of its higher sensitivity, is often the method of choice for glucose estimation – particularly in automated systems where cheap, stable reagents are vital. The NADP-linked assay, on the other hand, can be increased in sensitivity using fluorescence measurements and is very versatile in that it can be modified to assay a wide variety of substrates. It is the method of choice for measuring concentrations of the co-factors ATP and $NADP^+$ (see below).

7.4 **Modifications**

(a) *Fluorescent detection*

Use reaction mixture as above, except that the buffer should be filtered through a syringe-mounted Millipore (0.45 μm) filter before use to remove dust. Add 0.1 – 10 nmol of glucose sample. Read fluorescence with $\lambda_{ex} = 340$ nm, $\lambda_{emit} = 460$ nm. The cell holder should be thermostatted and calibration made with a known solution of $NADP^+$, whose concentration is determined spectroscopically on the same day.

(b) *To determine $NADP^+$*

Add 10 mM of glucose, 4 mM MgATP to the reaction mixture plus 40 – 400 nmol of sample (fluorimeter 0.1 – 10 nmol).

(c) *To determine ATP*

Add 10 mM glucose, 1 mM $NADP^+$ plus 40 – 400 nmol of ATP (fluorimeter 0.1 – 10 nmol)

In this case, increase in A_{340} might show a biphasic pattern (*Figure 3*). This is due to the slow oxidation of glucose itself by G6PdH. This is minimised by keeping the glucose concentration low (10 mM or less). To analyse the results, extrapolate the slow linear rise in A_{340} back to t=0, to calculate ΔA.

(d) *To determine other sugars*

Sucrose – include invertase.
Fructose – include hexose phosphate isomerase.
G1P – include glucose phosphate isomerase, etc.

By adding enzymes in a suitable order, a series of sugar metabolites in the same solution can be measured (*Figure 8*).

A variety of NAD^+ and $NADP^+$-linked assays are described in refs. 29 and 44.

8. RATE METHODS OF ASSAY

All assays given thus far are 'end point' methods, as is expected from the higher precision of this type of assay (Section 2.3). However, in some specific cases an assay depend-

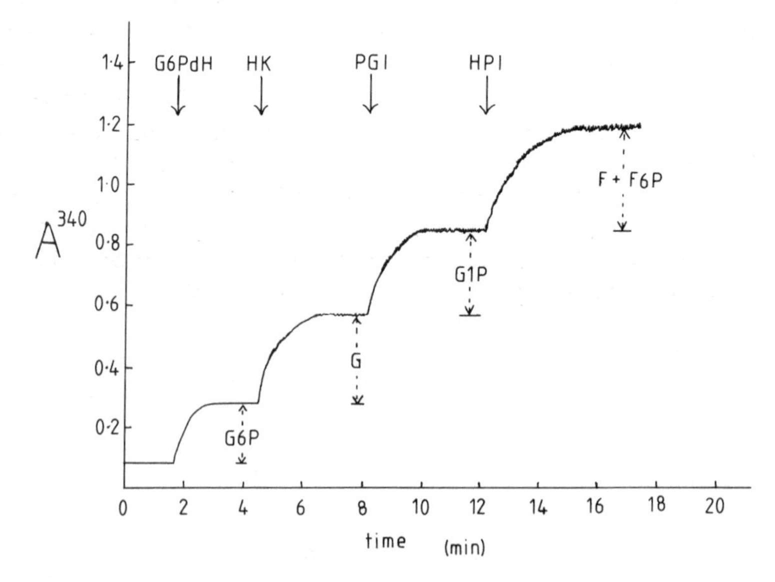

Figure 8. Sequential assay of hexose metabolites. The reaction mixture contained 200 mM Tris, 6 mM $MgCl_2$, 0.2 mM EDTA, 4 mM ATP, 1 mM $NADP^+$ at pH 8 (HCl). About 30 nmol each of glucose, glucose-1-phosphate, glucose-6-phosphate, fructose and fructose-6-phosphate were also present per ml of reaction medium. 1 – 2 U of the enzymes G6PdH, hexokinase, phosphoglucoisomerase and hexose phosphate isomerase were added as indicated, and the A_{340} followed.

ing on the rate of a given process – typically an enzyme reaction – may be useful. Where $s \ll K_m$, the rate (v) of an enzyme reaction is proportional to substrate concentration (s). Thus measurement of v can give, in some cases, an estimate of s. Calibration with known amounts of substrate, and careful control of temperature, ion composition of the buffer, etc., are vital to the precision of this type of assay, since enzyme rates are very sensitive to environmental conditions.

Rate assays are thus likely to be less precise than end point assays. They are also likely to be less sensitive, since measurements are made before all substrate is converted into product. They are used only in a few special cases, where they permit use of some very sensitive measuring method (e.g., bioluminescence) or where they permit 'amplification' – the production of several mols of product for each mol of material under assay (e.g., when this substance is a co-factor or allosteric effector).

8.1 'Luminescent' Rate Assays

In recent years, the availability of photometers (instruments which measure light output *without* an incident beam) has led to detailed investigation of measuring techniques using luminescent (light yielding) reactions. The sensitivity of such instruments is potentially much greater than that of spectrophotometers for the same reason as is the sensitivity of fluorimeters – we are measuring an increase in measured light from zero rather than a decrease in a strong incident beam. In principle, photometers should be the least complex of the three – requiring only a light-tight box, a stable photomultiplier and an injection point – but since few compounds are naturally luminescent, dedicated instruments are rarely found in the average laboratory. However, scintillation counters – often available for the measurement of radioactivity – can normally be simply modified for luminescent analysis, and a technique using a scintillation counter is described below (30).

Three types of bioluminescent assay may be useful in biochemical analyses:

(i) The detection of ATP using firefly luciferase (can be modified to detect ADP, GTP, ATP-producing agents such as creatine kinase, etc.) (31).

(ii) The detection of NADPH using bacterial luciferase (can be modified to detect NADH or any system linked to NADPH or NADH production) (32).

(iii) The detection of H_2O_2 using luminol (chemiluminescence) (can be used in oxidase based assays, measurement of phagocytosis, etc.) (33).

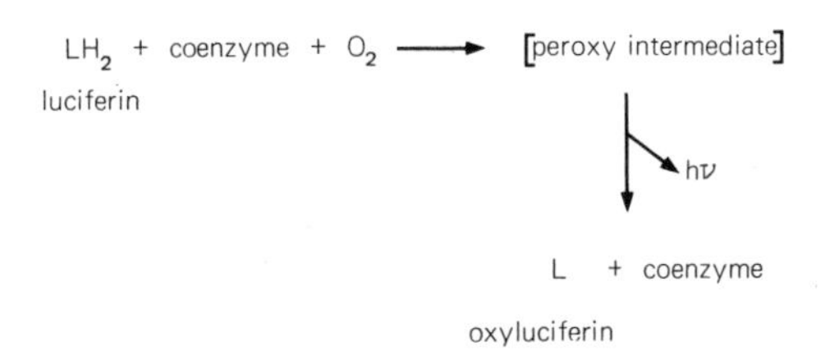

Figure 9. General reaction for bioluminescence. The luciferins are complex organic molecules (see Figure 10). The coenzymes are: ATP in firefly luciferase; FMN in bacterial luciferase; Ca^{2+} in *Aequorea* lucciferase ('aequorin'). In the last case, the peroxy intermediate is stored, bound to a protein, until Ca^{2+} is added.

In all cases, a complex organic compound (a 'luciferin', in enzyme-based systems) is oxidised to a peroxy compound, which breaks down to an excited oxy compound – and this is the species which emits light (*Figures 9* and *10*). (In principle the oxidation is so exergonic that the only way to remove the energy is by light emission.) In the two enzyme-based systems, oxidation is dependent on prior modification of the luciferin with the co-enzyme under assay.

8.2 **Determination of ATP Using Firefly Luciferase**

ATP reacts with firefly luciferin, to form adenyl-luciferin, which is oxidised by air with light emission. Both reactions are catalysed by firefly luciferase (*Figure 10*).

Conditions are chosen so that:

(i) [ATP] is well below its K_m, so that the rate of light emission is proportional to its concentration.
(ii) small amounts of enzyme are used, and competitive inhibitors of luciferase included, so that the amount of ATP used up in the reaction is negligible, and the light output is constant.

8.2.1 *Reagents*

(a) 0.1 M Tris, 10 mM $MgCl_2$, 2 mM EDTA, pH 7.75 (reaction buffer).
(b) 10 mg/ml of bovine serum albumin.
(c) 1 mM sodium pyrophosphate (pH 7.75).
(d) 10 μM or 1 μM ATP.
(e) 1 mg/ml of luciferin in reaction buffer.
(f) 1 mg/ml of luciferase plus 5 mg/ml of albumin in reaction buffer.

Solutions (b), (c), (e), (f) are stored at $-20°C$, but note that (i) the luciferin solution should be kept dark, by wrapping the tube in foil, and (ii) luciferase loses activity on freeze-thawing, and should be frozen in aliquots, to be thawed not more than twice

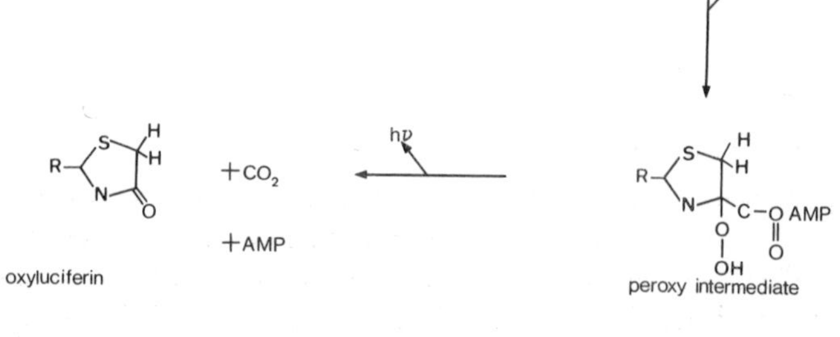

Figure 10. Mechanism of firefly luciferase reaction (46).

each. Solution (a) is stable at 4°C for months. Solution (d) is prepared fresh on day of use, from a stock of 1 mM ATP which is stored as aliquots at −20°C. The 1 mM ATP solution should be calibrated spectrophotometrically (see Section 7.2). Micromolar solutions of ATP are not stable to freezing.

A convenient mixture of solutions (a), (b), (c), (e) and (f) is available commercially from LKB Produkta.

8.2.2 *Method*

The reaction takes place in low K^+ glass scintillation vials (Beckman low ^{40}K vials for LS counting no. 22-16198) to minimise light emission.

(i) Pipette into each vial: 2.5 ml of reaction buffer, 0.2 ml of albumin, 10 μl of Na_4PP_i, 20 μl of luciferin and 0.25 ml of sample.

(ii) Place the vials in the belt of a scintillation counter (e.g., Beckman LS233, LKB 2510 Ultrobeta) and allow to equilibrate at room temperature for 5 min. The coincidence control of the counter is switched off/overridden. [Scintillation counters cut down background (shot) noise by counting pulses of light, and scoring only those pulses detected at two different photomultipliers in coincidence. Since one ATP molecule gives one photon, which can be detected by only one photomultiplier, coincidence should be eliminated to measure bioluminescence.] The first vial should be a blank. Set the counter to read low energy radiation (0−10%, or half ^3H channel).

(iii) Lower the first vial manually into the counting chamber, and switch the counter to automatic count, for two periods of 0.1 min.

(iv) On completion of the first period (i.e., when the teleprinter begins to print out), add luciferase (20 μl, or see below) by syringe/automatic pipette to the next sample in line, mix the vial and replace in the belt. The counter will in the meantime repeat count of the first vial, and move the next vial into position for counting.

(v) Repeat the procedure for all samples. In principle, the repeated operation of the counter is used to standardise the time between addition of enzyme and measurement of light emission.

Since, as in fluorescence techniques, calibration is arbitrary, the amount of luciferase added can be adjusted to give convenient rates of light emission (e.g., 500−10 000 counts in 0.1 min).

It is convenient to determine the dilution of luciferase to be used by initial calibration with 10^{-11} – 10^{-10} mol ATP, or 10^{-10} – 10^{-9} mol ATP (e.g., six different values in one of these ranges).

8.2.3 *Interference*

(a) *By small molecules and ions*

Many compounds (metallic cations, K^+, Ca^{2+}, etc., anions such as Pi, perchlorate, glycerol and other organic solvents) considerably affect the rate of the luciferase reaction. These effects are minimised by keeping the sample volume to less than 10% of the assay volume. In addition, an internal standardisation should be carried out, where samples containing (i) the unknown solution and (ii) the unknown solution plus a con-

Table 2. Luciferase Assay for ATP – Troubleshooting.

Problem	*Cause*	*Remedy*
Low light output in ATP standards	(1) Coincidence on	Turn off photomultiplier coincidence
	(2) Luciferin omitted	Add luciferin
	(3) Luciferase inactive	Use fresh luciferase (loses activity on freeze/thawing)
Low light output in internal standards	(1) Inhibitory material in sample (e.g., salt, EDTA, perchlorate, aldehydes)	Use less sample
	(2) Coloured material in sample	Bleach sample with H_2O_2 and destroy excess with catalase
Rapid decay of light output	(1) Loss of luciferase activity in assay medium	Use clean vials (normally due to detergent contamination)
	(2) Loss of ATP from assay medium	(i) Purify luciferase to remove contaminating enzymes (ii) Add inhibitor of luciferase (arsenate, pyrophosphate) to slow down ATP utilisation (iii) Pre-treat sample to remove contaminating enzymes
High light output		
(1) Vial absent	Unstable photomultiplier	Use more modern scintillation counter
(2) Vial present, but no buffer	(1) $^{40}K^+$ decay	Use low K^+ vials
	(2) Phosphorescence of glass	Use low K^+ vials and keep vials in dark before use
(3) In blank (luciferase, but no sample)	ATP contamination	
	(a) In dispensers	Use disposable micropipettes, or wash any liquid dispensers with 100 mM sodium pyrophosphate and water before use
	(b) In luciferase solution	Use purified luciferase, or allow firefly extract 2 h at room temperature to remove intrinsic ATP
	(c) By bodily fluids	Care in handling. Wear gloves

venient known amount of ATP are measured in tandem. This also allows for colour quenching, where the colour of the sample (or its turbidity) decreases the amount of light reaching the photomultiplier (see Section 4.2).

(b) *By other enzymes*

Besides ionic/solvent effects on luciferase itself, enzymes that affect ATP levels will also interfere. These may be present in the luciferase preparation itself, or in the extract used. In the second case, a common activity observed even in perchlorate extracts of tissue is adenylate kinase, and for storage such extracts should be treated with EDTA to abolish its activity or the ATP/ADP ratio observed may be distorted (34).

Crude luciferase extracts, such as may be obtained direct from firefly tails (e.g., Sigma XR2976), contain adenylate kinase, nucleotide diphosphate kinase, creatine kinase and some phosphatase activity. These can be removed in part by gel filtration (35), but commercially available, purified luciferase should be used in preference. Such enzymes may lead to the observation of a light 'flash' – a rapid decrease in light output with time – rather than a steady (<5% decline over 5 min) output. This can be detected by comparing the first and second 0.1 min values for the counts measured.

8.2.4 *Modifications*

To detect ADP (+ATP), add pyruvate kinase and phosphoenol pyruvate at a suitable interval before the luciferase, to convert ADP into ATP. To detect creatine phosphate, ADP and creatine kinase are added. Various other ATP-luciferase linked assays have been described (31,36). Note that the extreme sensitivity of the technique means that coupling enzymes should be extensively dialysed (or gel filtered) to remove any contaminating nucleotides, and ADP itself should be purified to remove ATP by ion exchange chromatography (31).

9. ASSAYS INCLUDING 'AMPLIFICATION'

9.1 Metabolite as Allosteric Effector

Amplification, the production of several molecules of compound A following the application of one molecule of compound X, is a well known phenomenon in biology. In its simplest case, compound X (a hormone, for example) is used to switch on an enzyme (adenyl cyclase, for example) to produce many molecules of product (cAMP in this case). This is useful both *in vivo*, in hormone action, or *in vitro*, for measurement of hormone by bioassay. Under standard conditions, for example, the concentration of X governs the rate of the enzyme reaction, and thus the amount of product A formed in a given time.

Note that, in this case, the production of A takes place (i) by an irreversible, downhill reaction (ATP→ cAMP + 2Pi), (ii) under conditions where the enzyme is saturated with its precursor, (ATP) but usually (iii) with excess of enzyme over compound X (hormone). Clearly, if the product A were itself coloured, or could be converted easily into a coloured compound, the concentration X could be measured by spectrophotometric measurement of A.

$$B \xrightarrow{X} A \qquad \text{Equation 7}$$

X in an allosteric effector of the reaction.

9.2 Metabolite as Regenerated Co-factor

The amplification principle has been adapted (29) for the measurement of compounds, such as co-factors, which can be used and re-synthesised in opposing chemical reactions. In this case, Equation 7 must be expanded

$$\begin{matrix} B \\ A \end{matrix} \circlearrowright \begin{matrix} X \\ X' \end{matrix} \circlearrowleft \begin{matrix} Q \\ P \end{matrix} \qquad B + P \xrightarrow{(X+X')} A + Q \qquad \text{Equation 8}$$

The rate of production of either A or Q (either or both of which may be measurable spectrophotometrically) is governed by X or X′ concentration, respectively, and in the steady-state [X] and [X′] are unchanged over the time of the reaction. Again, clearly, one molecule of X is re-used to produce many molecules of A, and this overcomes the inherent insensitivity of rate assays caused by incomplete reaction of the substance under assay (see above).

The same principles apply as in the case of the hormone-governed reaction – B and P are present in excess and at near-saturating levels, and each of the two part reactions is irreversible (ΔG^o large and negative). The rate of the cycle is dependent on the enzyme and on $(X+X')$, which is present far below saturating levels. It is also convenient to carry out the reaction with an excess of enzymes over $(X+X')$, so that essentially all the co-factor is bound and the cycling rate is relatively insensitive to the amount of enzymes actually present (increased precision – insulation from pipetting errors).

10. ASSAY OF ($NADP^+$ + NADPH) (29)

$NADP^+$ is used to oxidise excess G6P. The NADPH produced is re-oxidised using excess ketoglutarate and ammonia, and hence more G6P can be oxidised. The reaction is stopped after several cycles, and the product 6-phosphogluconate (6PG) assayed separately by an end point method (Section 7.2) using fluorescence detection for greatest sensitivity.

Reaction 1

G6P → 6PG (G6PdH); $NADP^+$ → NADPH (G6PdH); NADPH → $NADP^+$ (GludH); α-ketoglutarate + NH_3 → Glutamate (GludH)

Reaction 2 6PG + $NADP^+$ $\xrightarrow{\text{6PGdH}}$ ribose 5 phosphate + CO_2 + NADPH

The first step is a rate assay (rate depends on $NADP^+$ and NADPH levels), and needs to be carried out at regulated pH and temperature. After a suitable interval, this reaction is stopped, e.g., by heating (it is necessary to remove glutamate dehydrogenase or no net NADPH will be produced in the second stage), and 6PG is assayed in a spectrophotometer.

Since the first stage need not be carried out in a cuvette, and to increase reaction rate (amplification) for a given amount of enzyme used, the volume at this stage is kept small. Similarly, the first part of the reaction can be carried out under different conditions (optimal for its enzymes) from the second.

10.1 **Reagents**

PART I

(a) 100 mM Tris, 10 mM ammonium acetate, pH 8.0 (acetic acid).
(b) 100 mM α-ketoglutarate.
(c) 10 mM G6P.
(d) 1 mM ADP (activator of glutamate dehydrogenase).
(e) 100 U/ml of glutamate dehydrogenase.
(f) 100 U/ml of G6PdH.

Solutions I(b), (c), (d) are made up in 100 mM Tris acetate, pH 8.0, and stored at −20°C. The enzymes I(e) and I(f) are dialysed into 100 mM Tris acetate at 4°C, on the night before use, to remove all traces of NADPH, and of sulphate which inhibits the reaction.

PART II

(a) 40 mM Tris, 30 mM KCl, 5 mM $MgCl_2$, 0.1 mM EDTA, pH 8.0 (HCl).
(b) 10 mM $NADP^+$ (pH 6.5).
(c) 10 U/ml of 6PG dehydrogenase [in buffer II(a)].

Solution II(b) is stored frozen at −20°C. Solution II(c) is made by diluting stock enzyme solution into buffer II(a) on day of use.

10.2 Method

(i) Prepare a stock solution containing solutions I(a) – I(f), mixing the solutions in the volume ratios 80(a):5(b):10(c):10(d):10(e):10(f). (Preparing a stock solution cuts down variability between samples.)
(ii) Add 125 μl of this mixture to a microtest tube, and incubate for 5 min at 37°C.
(iii) Start the reaction by adding 20 μl of sample solution [$10^{-13}-10^{-12}$ mol NADP(H)] and incubate for 1 h.
(iv) Stop the reaction by heating at 100°C for 2 min.
(v) Transfer the contents of the microtest tube quantitatively to a fluorimeter cuvette by washing with 4 × 0.5 ml of buffer II(a).
(vi) Add a final 0.3 ml of buffer II(a) and 0.1 ml of $NADP^+$ (10 mM).
(vii) Measure the fluorescence. (λ_{ex} = 340 nm, λ_{emit} = 460 nm.)
(viii) Add 0.1 ml of 6PG dehydrogenase and follow the reaction to completion (Section 7.2).
(ix) Prepare a blank with no added NADP(H) [at point (iii)].

Calibration requires preparation, on the day of assay, from a stock of 10 mM $NADP^+$ (which has been accurately calibrated by spectrophotometry) a dilute $NADP^+$ solution, suitable for dispensing samples of $10^{-13}-10^{-12}$ mol $NADP^+$. Calibration of the fluorimeter response to NADPH, although not essential, may be useful.

As in all highly sensitive assays, scrupulously clean glassware and care in handling materials are essential. All enzymes should be dialysed into the reaction buffer extensively before use. It is also necessary either to use ultrapure reagents, or to chromatograph (ion exchange) phosphorylated reagents before use – ADP commonly contains traces of $NAD(P)^+$ and ATP, and *vice versa*. A rapid and convenient chromatographic system for ADP purification has been described (31).

10.3 Modifications

Since $NADP^+$ and NADPH are interconverted above, the above method determines the sum of both forms. It is possible to assay specifically one of these two forms by first destroying the other.

(a) *To determine NADPH.* Treat the sample (20 μl) with 5 μl of 0.5 M NaOH at 80°C for 20 min to destroy $NADP^+$. Neutralise the solution with 5 μl of 0.5 M Hepes (acid form) and then carry out the cycling procedure as above.

(b) *To determine $NADP^+$.* Treat the sample (20 μl) with 5 μl of 0.05 M HCl containing 0.15 M ascorbic acid. [The ascorbic acid protects the NAD(P)H against oxidation by air, which would lead to artifactual increases in $NAD(P)^+$ levels.] Heat at 60°C for 30 min, cool in ice and neutralise by the addition of 5 μl of Tris base. Carry out the cycling procedure as above.

(c) *To determine other co-enzymes.* NAD^+/NADH can be recycled using lactate dehydrogenase/glutamate dehydrogenase, and can be assayed independently by destroying the other form as above. ATP/ADP can be recycled using hexokinase/pyruvate kinase.

In either case, sensitivity can be increased by using a luminescent assay for ATP or NADH (see above).

11. SPECTROPHOTOMETRIC MEASUREMENT OF ENZYME ACTIVITIES

11.1 Introduction

The rate of enzyme action (v) is the quantity of substrate it converts into product per unit time (μmol/min). This varies with substrate concentration s, unless $s \gg K_m$ (conveniently $s > 10\ K_m$), when v is close to the maximal velocity, V_m.

Measurements of v at different values of s allow elucidation of K_m and V_m for the amount of enzyme used. K_m is independent of enzyme concentration, and has units of concentration (M), representing s when $v = 1/2\ V_m$ in the absence of enzyme inhibitors. V_m is dependent on enzyme concentration and may be in a variety of units.

(a) V_m in μmol/min/mg protein (U/mg). V_m represents the specific activity and increases with enzyme purity.

(b) V_m in μmol/min/mg tissue. V_m represents the maximal capacity of the tissue for the reaction in question.

(c) V_m in mol/min/mol enzyme (min^{-1}). V_m represents the turnover number of the enzyme.

11.2 Spectrophotometric Measurement

If the absorption or fluorescence of substrate, S, differs from that of product, P, the activity of an enzyme can be measured directly by following the change in absorbtion (fluorescence) in time. This situation applies to all dehydrogenases using $NAD(P)^+$ as co-enzyme, and to some other reactions where changes in double bond configuration are involved (uricase, fumarase, etc.). An assay where the colour change is followed (normally by a chart recorder) as the reaction proceeds is known as a continuous enzyme assay. Note that the spectrophotometer measures $\frac{dA}{dt}$, the rate of absorbance change, which is proportional to $\frac{ds}{dt}$, the change in concentration with time. To measure the amount of substrate converted, the volume of the mixture (x) needs to be taken into account

$$v = \frac{ds}{dt} = x.\frac{ds}{dt} = \frac{x}{\epsilon_\lambda} \cdot \frac{dA}{dt} \text{ mol/min} \qquad \text{Equation 9}$$

where ϵ_λ is the molar absorptivity, x is the volume of assay mixture in litres, and time, t, is measured in minutes.

If no convenient colour change occurs as the reaction proceeds, we must adopt one of the following approaches for spectrophotometric measurement.

(i) Stop the reaction, conveniently with acid, alkali or organic solvent, and add a reagent (chromogen) which produces a colour on reaction with one of the products. This is termed a *discontinuous assay*.

(ii) Couple the reaction under study with an indicator reaction (usually, but not necessarily, enzymic), which produces a colour 'instantaneously'. Thus the production of G6P from glucose can be monitored directly if large amounts of G6PdH and $NADP^+$ are present, since NADPH – with a measurable absorbtion – is rapidly formed (see Section 7.2). This is a *coupled assay*.

(iii) Use, instead of the natural substrate, a synthetic chromogenic substrate. For example, phosphatase activity can be monitored by its ability to release nitrophenolate anion (yellow) from nitrophenol phosphate. This method cannot give a good indication of rates with the normal substrate, unless separately calibrated, but has proved very useful in the study of proteases and other hydrolases where the true substrate is complex and little different chemically from the product.

11.3 **Requirements of Coupled Assay Systems**

In a coupled assay, the product (colourless) of the reaction under study is used instantaneously to produce a coloured compound in a secondary, 'indicator', reaction. Many biological assays conveniently use a dehydrogenase as indicator reaction, since

(a) NAD(P)H has a high extinction at 340 nm, and
(b) $NAD(P)^+$ and NAD(P)H are unlikely to affect (poison) the primary reaction.

$$\text{Primary reaction } AH_2 \rightleftharpoons BH_2 \qquad \text{rate } v_a \qquad \text{Equation 10}$$

$$\text{Indicator reaction } BH_2 + NAD^+ \rightarrow B + NADH + H^+ \qquad \text{rate } v_b \qquad \text{Equation 11}$$

This procedure is analogous to that described for metabolite assays (Section 7.2). However, if the activity of the primary enzyme is to be monitored, it is essential that the rate of NADH production (in Equation 11) should equal the rate of the primary reaction – the two reactions must be *coupled* together. If ΔG for the dehydrogenase is large and negative (use a trapping system if necessary), this requirement simply means that the indicator enzyme(s) must be present in sufficiently large amounts so as not to limit the primary reaction.

How much of the indicator enzyme(s) (enz_b) should we add? Clearly, its maximum activity should be well above that of the measured enzyme (enz_a). If this is the case, the intermediate (BH_2 above) is kept low and enz_b works far below its K_m, and thus

$$v_b = \frac{V_m^B}{K_m^B}[BH_2].$$

Initially, however, $[BH_2] = 0$, and the rate of the indicator reaction will lag behind that of the primary reaction (see *Figure 11*). By an analysis similar to that in Section 6.3, the time taken for $[BH_2]$ to approach 99% of the maximal value (and thus v_B to reach 99% of its maximal rate, v_A) is given by

$$t = 4.6\frac{K_m^B}{V_m^B}.$$

Whereas in Section 6.3 we were able to allow the enzyme 5 – 10 min to reach 99% completion, for kinetic measurements we need the approach to 99% maximal rate to

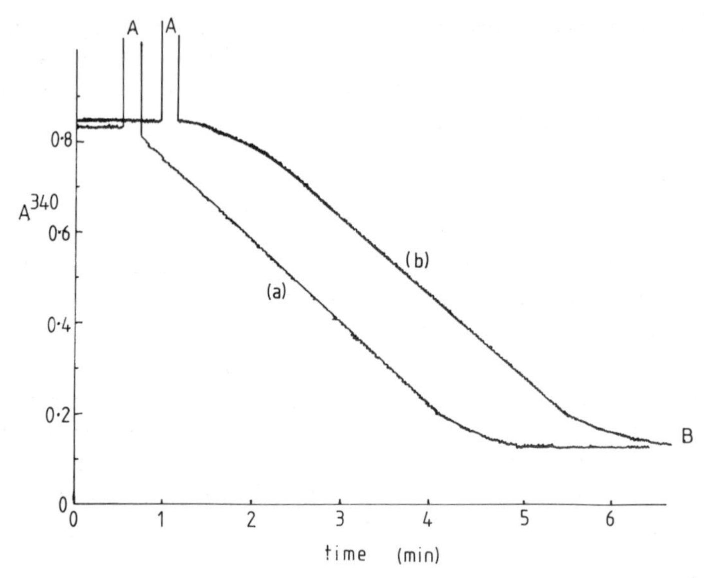

Figure 11. Effects of insufficient indicator enzyme on coupled ATPase assay. The reaction mixture contained 100 mM Tris, 50 mM KCl, 6 mM $MgCl_2$, 4 mM ATP, 1 mM phosphoenol pyruvate, 0.2 mM EDTA, 0.15 mM NADH and 10 U/ml pyruvate kinase (pH 8.0, HCl). In addition, 10 U/ml (**a**) or 0.5 U/ml (**b**) of lactate dehydrogenase was present. The reaction was started by addition of 0.03 U/ml of purified mitochondrial ATPase (F_1).

be virtually instantaneous. A suitable lag time is less than 10% of the measuring period; say, 0.1 min on a 2 min measuring period. Assuming K_m^B for the indicator reaction is around 1 mM

$$t = 0.1 = \frac{4.6 \times 1}{V_m^B}$$

or

$$V_m^B = 46 \text{ U/ml reaction medium}$$

If V_m^B is less than this, the lag time will increase – if, say, $V_m^B = 5$ U/ml and $K_m =$ 1 mM, the rate of the indicator reaction will not attain 99% of the rate of the primary reaction over a 2 min assay period. Clearly, indicator enzymes with a low K_m are preferred.

In parallel with Section 6.3, the amount of indicator enzyme is apparently independent of the amount of primary enzyme. This may at first sight appear surprising but the reader should note that, for this analysis to be correct, $[BH_2] \ll K_m^B$, i.e., $v_B \ll 1/10\ V_m^B$, so that V_m^A also must be less than $1/10\ V_m^B$. A more detailed analysis of this system has been published (37).

The activity should be measured under conditions optimal for the primary reaction, since this is the one under study. For many reactions this may be close to the optimum for the indicator enzymes, but this is by no means always the case. In most cases, this can be overcome by adding more of the coupling enzymes (i.e., noting that V_m^B above refers to rates in the assay buffer used, not necessarily the values given on the bottle), but in some cases the two systems may be incompatible. It is not possible to measure

Ca^{2+} ATPase or methanol-stimulated ATPase by the coupled assay system in Section 12.2, for example, since Mg^{2+} is required by, and methanol leads to inactivation of, the coupling enzymes. The discontinuous assay (Section 12.1) where reaction and colour development are separate, must then be used.

11.4 **Linearity**

For any of these cases, the assay should show

(i) Linearity in time over the period of assay, i.e., measurement over x and 2x min of reaction should give responses Δ and 2Δ.

(ii) Linearity with enzyme concentration, i.e., measurement with *e* and 2*e* mg of enzyme should give responses Δ' and $2\Delta'$ in a given time.

In setting up any assay, these two conditions should be verified by direct measurement and, if not satisfied, the assay must be redesigned (less enzyme, inclusion of product trap, etc.). In any assay series, it is always better to include two different concentrations of enzyme, to check linearity and internal consistency, in place of two identical duplicates (although duplicates are of course valuable in themselves as measurements of assay precision).

12. MEASUREMENT OF ATPase ACTIVITY

ATPase activity can be measured using either a discontinuous or a continuous assay method (see below). A chromogenic substrate has also been reported (38). Its measurement, while widely used in itself, also demonstrates several points to be observed in the design of enzyme assays in general.

12.1 **Discontinuous Assay**

The enzyme is allowed to hydrolyse ATP for a given time, after which the reaction is quenched with acid. Acid molybdate reagent is added to complex the inorganic phosphate formed, and the colour of the phosphomolybdate measured after enhancement by reduction, or by extraction into organic solvent. The method described determines V_m for enzymes where K_m is less than 500 μM.

12.1.1 *Reagents*

(a) 100 mM Tris, 50 mM KCl, 2 mM $MgCl_2$, 0.2 mM EDTA, pH 8.0.

(b) 200 mM MgATP (pH 7.0) (see Section 7.2).

(c) 40% w/v TCA (stopping reagent).

(d) 1% w/v ammonium molybdate in 1 M H_2SO_4 (acid molybdate).

Solutions (c) and (d) are stable at room temperature for months. Solution (b) is stored frozen at −20°C, in aliquots, and frozen/thawed no more than twice before discarding.

The colour reagent is prepared, on the day of use, by dissolving 1 g of $FeSO_4.7H_2O$ in acid molybdate solution (d) (39).

12.1.2 *Method*

(i) Prepare 0.8 ml of reaction buffer (a) and 0.2 ml of sample (0.02−0.1 U ATPase).

(ii) Incubate at 30°C for 5 min.
(iii) Start the reaction with 20 μl of MgATP (200 mM).
(iv) After 10 min, stop the reaction with 0.1 ml of 40% TCA.
(v) Add 1.9 ml of colour reagent, and read the A_{700} after 10 min. [Note that the reduction of phosphomolybdate to a blue complex was integral to the Lowry protein assay (Section 3.4) but in this former case the phosphomolybdate was supplied preformed and the protein used as the reducing agent.] The colour is stable for a further 10 min, but slowly darkens due to non-enzymic hydrolysis of ATP in the acid.
(vi) 1 μmol Pi released gives an A_{700} ~1.

Blanks should be prepared with no enzyme or, better, with a boiled enzyme solution, since the enzyme solution itself may contain traces of phosphate (e.g., phosphate buffer).

A calibration line should be prepared using 0.1–0.8 μmol Pi, which is added from a standard solution of 10 mM NaH_2PO_4 solution. This is dissolved, not in water, but in 1% TCA to prevent bacterial growth, and can then be stored at 4°C indefinitely.

Organic amines (triethylamine, cyclohexylamine) and neutral detergents (Triton) interfere (see Lowry protein assay, Section 3.4). The effect of neutral detergents may be overcome using the organic solvent procedures below, although it may also be necessary to increase molybdate concentration since some complexation of these reagents with the molybdate does occur. Some divalent metal ions (e.g. Cu^{2+}) catalyse a non-enzymic hydrolysis of ATP.

12.1.3 *Notes*

(i) Since many common detergents contain phosphate, it is convenient to wash the tubes for this assay in 6 M HCl, rinse and dry, and to use these tubes for no other purpose.
(ii) The reaction is started by addition of MgATP, since this is likely to be the most labile species in the assay medium. Furthermore, a small error in its addition volume (as may occur when starting at timed intervals) will hardly affect the reaction rate since MgATP should be present at near-saturating levels. Small variations in enzyme volume added, by contrast, will affect the reaction rates observed, decreasing precision. Initiation with substrate is a common feature of enzyme assays where V_m (specific activity) is to be measured.
(iii) Other reducing agents ($SnCl_2$, ascorbic acid) may replace ferrous sulphate.

12.1.4 *Modifications*

(a) *Product inhibition*

If the ATPase is subject to inhibition by ADP, the ADP should be converted back to ATP by a regenerating system. This has the additional advantage that ATP levels do not change during the assay. (Product inhibition is indicated if the reaction is non-linear in time and substrate depletion over the course of the reaction is >5%.)

In this case, the volume of reaction buffer and enzyme is reduced to 0.9 ml and 1 U of pyruvate kinase is included in it. The reaction is initiated by a mixture of MgATP and phosphoenol pyruvate [0.1 ml 40 mM MgATP/20 mM phosphoenol pyruvate (pH 7.0)] which can be prepared in advance and stored frozen in aliquots.

(b) *Turbidity*

If more than about 20 μg of protein is used in this assay (for example, if a membrane-bound ATPase is measured), addition of TCA may lead to cloudiness. This can be overcome

(i) by centrifuging the solution on a bench centrifuge, followed by removal of a sample (0.6 ml) to the colour reagent;
(ii) by addition of acid molybdate (without $FeSO_4$) together with an organic solvent to stop the reaction (40), in place of TCA. In this case, mix 1 ml reaction mixture with 4 ml of acid molybdate (d): acetone in 1:1 ratio. Add 0.4 ml of 1 M citrate and read the A_{355}. [All volumes in this assay can be conveniently reduced to one half.] The yellow colour of the phosphomolybdate complex is enhanced by the organic solvent and no reduction is necessary. 1 μmol Pi gives an A_{355} $\sim$3. If the solution is still turbid, 8 mg/ml of SDS may be added to the molybdate/acetone mixture to dissolve protein.

In this modification, citrate is present to complex excess molybdate. This has the advantage that any phosphate released (by acid-catalysed hydrolysis) after adding the stopping reagent has no molybdate to complex with, and the colour is stable for several hours (compare above). However, it is important not to add citrate before the molybdate reagent.

Reducing agents interfere with this method. If the solution begins to turn green or blue, a drop of H_2O_2 (40%) should be added immediately after the molybdate/acetone mixture (see Section 3.4).

Citrate, in similar concentration, can be used to prevent detection of non-enzymatic hydrolysis in the reducing assay above. If it is used, however, ascorbic acid (2%) should be used as reductant, since $FeSO_4$ is precipitated as iron citrate.

(c) *Contaminating phosphatases*

Crude ATPase preparations may contain other enzymes – in particular phosphatases – capable of releasing Pi from organic phosphates. To confirm that the activity measured is due to the enzyme under study:

(i) A blank should be prepared in the absence of Mg^{2+}. All ATPases, but few phosphatases, require Mg^{2+}.
(ii) The reaction should be performed both with and without a specific inhibitor of the ATPase under study (ouabain for the NaK^+ ATPase, oligomycin for the mitochondrial ATPase, etc.) if one is available.

(d) *Increased sensitivity*

The sensitivity (ability to measure smaller amounts of ATPase) of the above assays can be increased simply by carrying out the reaction for longer times before quenching. However, this approach is limited, first by the patience of the operator, but more importantly by the non-enzymic hydrolysis of the high energy phosphates in the assay solution over long periods of time. In general, incubation times of more than 60 min (detection of <5 mU ATPase) are impractical.

Sensitivity can be improved, however, by using a more sensitive method of phosphate determination, providing that:

(i) tubes are kept scrupulously clean (see above);
(ii) non-specific hydrolysis over the period of the assay, and phosphate in the initial reagents – in particular ATP and phosphoenol pyruvate – are negligible. Use of high quality reagents, careful pH adjustment of the solutions and the rejection of reagents after freeze/thawing (store in aliquots) is essential.

In the method given below (42) sensitivity is increased some 10-fold (detection of 2–20 mU ATPase) using the extraction of phosphate into butyl acetate as an enhancing method.

(i) Prepare the reaction mixture as above, using 2–20 mU ATPase.
(ii) Stop the reaction by addition of 0.2 ml of perchloric acid (1 M).
(iii) Add 0.1 ml of imidazole (250 mM, pH 5), 0.5 ml of 1.5% w/v sodium molybdate in water and 2.5 ml of butyl acetate.
(iv) Mix for 15 sec vigorously using a vortex mixture and allow 30 sec for the layers to separate (or spin gently on a bench centrifuge for 2 min).
(v) Remove the upper (butyl acetate) layer and read the A_{310}.
(vi) 0.1 μmol Pi gives A_{310} ~1.

If more than 20 μg protein is present, centrifugation may be required after addition of perchloric acid. If reducing agents are present, add one drop of H_2O_2 (40%) before the sodium molybdate (see above). Non-specific hydrolysis is minimised here, since Pi is immediately removed from organic phosphates by the extraction procedure.

Even more sensitive methods of Pi determination than this are available, notably one in which malachite green is used to enhance the colour of the phosphomolybdate (70) but they are rarely useful for ATPase determination due to problems with background phosphate. Radioactive methods, using [^{32}P]ATP, are then the method of choice (41).

12.2 **Coupled ATPase Assay (45)**

ATP is hydrolysed to ADP + Pi in the primary reaction. ADP is converted back to ATP using phosphoenol pyruvate and pyruvate kinase (thus, incidentally, removing the product inhibitor ADP and keeping [ATP] constant), and the pyruvate produced is reduced with lactate dehydrogenase. A_{340} declines.

lactate) LDH (NAD^+
ATP pyruvate NADH
ATPase ( ) PK (
ADP PEP

12.2.1 *Reagents*

(a) 100 mM Tris, 50 mM KCl, 2 mM $MgCl_2$, 0.2 mM EDTA, pH 8.0 (HCl).
(b) 200 mM MgATP (see Section 7.2).
(c) 50 mM phosphoenol pyruvate pH 7.0 (dissolve the monopotassium salt of phosphoenol pyruvate in water, and neutralise carefully with 1 M KOH).
(d) 100 mM NADH (in Tris buffer, pH 7.5) (see Section 7.2).
(e) 1000 U/ml of lactate dehydrogenase (in ammonium sulphate supsension).
(f) 2000 U/ml of pyruvate kinase (in ammonium sulphate suspension).

Solutions (b), (c), (d) are stored frozen at −20°C, and suspensions (e) and (f) at 4°C.

These solutions are conveniently mixed into a stock assay buffer on the day of use. Suitable volumes are 10 ml (a), 200 μl (b), 200 μl (c) and 20 μl each of (d), (e) and (f). The assay buffer is stored on ice. The A_{340} should be about 0.9 and can be adjusted by adding more or less NADH.

12.2.3 *Method*

A thermostatted sample holder in the spectrophotometer is required.

(i) Add 2.5 ml of the reaction buffer to a cuvette and incubate at 30°C.

(ii) When the temperature has equilibrated, add 0.05−0.2 U ATPase, mix well.

(iii) Measure the decline in A_{340} with time. From $\frac{\Delta A}{\Delta t}$, the rate $\frac{ds}{dt}$ can be calculated (Section 3.1). $A_{340}^{1\ \mathrm{mM}}$ for NAD(P)H = 6.22.

To avoid contamination between sequential samples, it is convenient to wash the cuvettes with acetone and dry well with warm air between samples.

12.2.4 *Modifications*

(a) *To eliminate other enzyme reactions*

As in Section 12.1 above, attempts should be made to ensure the measurement is specific for the enzyme to be measured, e.g., by testing the system with and without a specific inhibitor of this enzyme. In an NADH-linked assay, oxidases in the sample (for example, in the mitochondrial membrane) can oxidise NADH independently of the ATPase being

Table 3. Coupled NAD^+-linked Assay for ATPase Activity − Troubleshooting.

Problem	*Cause*	*Remedy*
Drift in A_{340} before addition of ATPase	Contamination with ATPase	Wash cuvettes with acetone between samples and dry well
	Bacterial contamination in buffer (cloudiness in buffer)	Make up new reaction buffer
A_{340} <0.05 before ATPase addition	Insufficient NADH present	Add more NADH
	ATP or phosphoenol pyruvate hydrolysed in stock solutions	Prepare fresh ATP or phosphoenol pyruvate
No reaction observed	Component missing	(1) Add ADP to check response of system (~1 μmol)
	Component inactive	(2) Add pyruvate to check response of lactate dehydrogenase (1 μmol)
		(3) Add each component individually, with ADP present, to find missing component, and replace in full medium
	Inactivation of enzymes due to carry over of organic solvents	(4) Repeat with clean, dry cuvette
Initial rapid drop	ADP added with enzyme	Ensure enough NADH (and other components) so that linear rate is observed
Initial lag	Coupling enzyme(s) not active enough (see Section 11.3)	Locate enzyme involved (normally LDH) as above, and replace at correct activity.

measured. Effects of this type can be detected by adding the ATPase to a reaction mixture containing all reagents but ATP. Addition of 1 mM KCN to the reaction medium will inhibit oxidases in many cases.

(b) *To increase sensitivity*

The sensitivity of this assay can be increased about 10-fold (detection limit 1 – 5 mU ATPase) using fluorimetric measurement, or using a spectrophotometer with f.s.d. = 0.2 absorbance. Above this sensitivity, the slow, non-enzymatic hydrolysis of phosphoenol pyruvate (and/or ATP) causes slow oxidation of NADH at rates comparable with the enzymic hydrolysis.

(c) *NAD(P)H-linked assays of other enzymes*

The use of coupled systems to measure enzyme reaction rates is a very versatile technique [as is the use of NAD(P)H-linked systems in metabolite assays – see Section 7.2]. A few examples are shown in *Table 4*. The reader is referred to Bergmeyer (44) for further examples.

Table 4. NAD(P)-linked Systems for Coupled Enzyme Assays.

Primary enzyme	*Coupling enzyme*	*Net reaction*
Kinase	PK, LDH	X+ATP→X-P + ADP ADP+PEP+NADH→ATP+lactate+NAD$^+$
Aldolase	TIM, glycerol Ⓟ dHase	Hexose diphosphate →2 triose Ⓟ triose Ⓟ + NADH→glycerol 3Ⓟ + NAD$^+$
α,β-Glucosidase	HK, G6PdH	α,β-Glucoside→glucose Glucose + ATP + NADP$^+$→ gluconate 6P + NADPH + ADP[a]
Citrate lyase	Malate dHase	Citrate→OA + acetate OA + NADH→malate + NAD$^+$
Glutamate-oxaloacetate transaminase	Malate dHase	α KG + asp→glu + OA OA + NADH→malate + NAD$^+$
Phosphoglycerate mutase	Enolase, PK, LDH	3-Phosphoglycerate→2PGA 2PGA+ADP+NADH→lactate+ATP+NAD$^+$
Phospholipase C	Diglyceride lipase glycerokinase glycerol Ⓟ dehydrogenase	Phospholipid→diglyceride + X-Ⓟ diglyceride+ATP+NAD→dihydroxyacetoneⓅ + FA + ADP + NADH
Succinyl CoA synthetase	PK, LDH	GTP+CoA+succ→succ CoA + GDP + Pi GDP+PEP+NADH→GTP+lactate + NAD$^+$[b]

[a]Either G6P or ADP could be assayed in this case. G6P is chosen as it is easier to measure an increase in A_{340} (F_{460}) from zero than a decrease from a high value.

[b]GTP reacts in the pyruvate kinase reaction as does ATP.

Abbreviations: PK, pyruvate kinase; LDH, lactate dehydrogenase; TIM, triose phosphate isomerase; HK, hexokinase; OA, oxaloacetate; 2PGA, 2-phosphoglycerate; FA, fatty acid; αKG, α-ketoglutarate.

13. CONCLUSIONS

This chapter describes the use of photometric methods in biochemical measurement – in the assay of macromolecules in general, in the assay of specific metabolites and in the measurement of enzyme activities. Employment of such techniques has been essential to the development of biochemistry as a quantitative science – in determining the nature of cellular components, in the elucidation of metabolic pathways and their control, etc.

The assays chosen as examples are among the most commonly used in biochemical laboratories. Through them the conflicting desires for simplicity and specificity, for reproducibility and sensitivity and for uniformity and versatility are illustrated. Among the themes which have emerged are the use of enzymes to confer specificity on rather general colour reactions, the recurrent use of pyridine nucleotide-coupled systems (in spectrophotometric, fluorimetric and luminometric measurements) and the principles behind the choice of quantities of reagents used. Thus, while this chapter cannot hope to give an exhaustive coverage of photometric assay techniques (see e.g., refs. 29,44), by use of the principles given the reader should be able to devise, or decide on, a spectrophotometric assay for almost any compound required.

14. REFERENCES

1. Rickwood,D., ed. (1984) *Centrifugation – A Practical Approach,* Second edition, published by IRL Press, Oxford.
2. Waters Column, **4**, no.2 p.8, Waters Associates, Northwich, Cheshire.
3. Clayton,R. (1963) in *Bacterial Photosynthesis,* Gest,H., San Pietro,A. and Vernon,L.P. (eds.), Antioch Press, Yellow Springs, USA, p. 495.
4. Lee,R.T., Denburg,J.L. and McElroy,W.D. (1970) *Arch. Biochem. Biophys,* **141**, 38.
5. Whatley,F.R. and Arnon,D.I. (1963) in *Methods in Enzymology,* Vol. 6, Colowick,S.P. and Kaplan,N.O. (eds.) Academic Press, p.308.
6. Buttner,H., Hansert,E. and Stamm,D. (1974) in *Methods of Enzymatic Analysis,* Bergmeyer,H.U. (ed.), Vol. 1, 2nd Engl. Edition, Verlag Chemie, p. 362.
7. Barbehenn,E.K., Law,M.M.Y., Brown,J.G. and Lowry,O.H. (1976) *Anal. Biochem.,* **70**, 554.
8. McKnight,G.S. (1977) *Anal. Biochem.,* **78**, 86.
9. Thorne,C.J.R. (1978) *Techniques in Protein and Enzyme Biochemistry,* Vol. **B104**, published by Elsevier, North Holland.
10. Foster,J.E. and Sternman,M.D. (1956) *J. Am. Chem. Soc.,* **78**, 3656.
11. Habeeb,A.F.S.A. (1972) in *Methods in Enzymology,* Vol. **25**, Hirs,C.M.W and Timasheff,S.N. (eds.), Academic Press, p. 457.
12. Miles,E.W. (1977) in *Methods in Enzymology,* Vol. **47**, Hirs,C.M.W. and Timasheff,S.N. (eds.), Academic Press, p. 431.
13. Spande,T.F. and Witkop,B. (1967) in *Methods in Enzymology,* Vol. **11**, Hirs,C.M.W. (ed.), Academic Press, p. 498.
14. Gornall,A.C., Bardawill,C.J. and David,M.M. (1949) *J. Biol. Chem.,* **177**, 751.
15. Cleland,K.W. and Slater,E.C. (1953) *Biochem. J.,* **53**, 547.
16. Warburg,O. and Christian,W. (1941) *Biochem. Z.,* **310**, 384.
17. Lowry,O.H., Rosebrough,N.J., Farr,A.L. and Randall,R.J. (1951) *J. Biol. Chem.,* **193**, 265.
18. Bensadoun,A. and Weinstein,D. (1976) *Anal. Biochem.,* **70**, 241.
19. Smith,P.K., Krohn,R.I., Hermanson,C.T., Mallia,A.K., Gartner,F.H., Provenzano,M.D., Mujimoto, E.K., Goeke,N.M., Olson,B.J. and Klenk,D.C. (1985) *Anal. Biochem.,* **150**, 76.
20. Bradford,M.M. (1976) *Anal. Biochem.,* **82**, 327.
21. Read,S.M. and Northcote,D.H. (1981) *Anal. Biochem.,* **116**, 53.
22. Udenfried,S., Stein,S., Bohlen,P., Dairman,W., Leimgruber,W. and Wiegele,M. (1972) *Science (Wash.),* **178**, 871.
23. Labarca,C. and Paigen,K. (1979) *Anal. Biochem.,* **102**, 344.
24. Karsten,V. and Wallenberger,A. (1972) *Anal. Biochem.,* **77**, 464.

25. Weiner,S., Weiner,R., Urivetsky,M. and Meilman,E. (1974) *Anal. Biochem.*, **59**, 489.
26. Greenberg,D.M. and Rothstein,M. (1957) in *Methods in Enzymology,* Vol. **4**, Colowick,S.P. and Kaplan,N.O. (eds.), Academic Press, p.708.
27. Barker,S.A., Somers,P.J. and Eptom,R. (1968) *Carbohydrate Res.*, **8**, 491.
28. Penefsky,H.S. (1977) *J. Biol. Chem.*, **252**, 2891.
29. Lowry,O.H. and Passoneau,J.V. (1972) *A Flexible System of Enzymatic Analysis,* published by Academic Press, NY.
30. Stanley,P.E. and Williams,S.G. (1969) *Anal. Biochem.*, **29**, 381.
31. Lundin,A. (1978) in *Methods in Enzymology,* Vol. **57**, de Luca,M. (ed.), Academic Press, p. 56.
32. Stanley,P.E. (1978) in *Methods in Enzymology,* Vol. **57**, de Luca,M. (ed.), Academic Press, p. 215.
33. Seitz,W.R. (1978) in *Methods in Enzymology,* Vol. **57**, de Luca,M. (ed.), Academic Press, p. 445.
34. Harris,D.A. and Slater,E.C. (1975) *Biochim. Biophys. Acta,* **387**, 335.
35. Nielson,R. and Rasmussen,H. (1968) *Acta Chem. Scand.*, **22**, 1757.
36. Karl,P.M. (1978) in *Methods in Enzymology,* Vol. **57**, de Luca,M. (ed.), Academic Press, p. 85.
37. Easterby,J.S. (1981) *Biochem. J.*, **199**, 155.
38. Kagawa,H., Kuwajima,T. and Asai,H. (1974) *Biochim. Biophys. Acta,* **338**, 496.
39. Sumner,J.B. (1944) *Science (Wash.),* **100**, 413.
40. Heinonen,J.K. and Lahti,R.J. (1981) *Anal. Biochem.*, **113**, 313.
41. Nelson,N. (1980) in *Methods in Enzymology,* Vol. **69**, Academic Press, p. 301.
42. Jenkins,W.T. and Marshall,M.M. (1984) *Anal. Biochem.*, **141**, 155.
43. Lanzetta,P.A., Alvarez,L.J., Reinach,P.S. and Candia,O.A. (1979) *Anal. Biochem.*, **100**, 95.
44. Bergmeyer,H.U. (1974) *Methods of Enzymatic Analysis,* 2nd Engl. Edition, Vols. **3**,**4**, published by Verlag Chemie.
45. Rosing,J., Harris,D.A., Kemp,A.Jr. and Slater,E.C. (1975) *Biochim. Biophys. Acta,* **376**, 13.
46. Tal,M., Silberstein,A. and Nusser,E. (1985) *J. Biol. Chem.*, **260**, 9976.

CHAPTER 4

Measurement of Ligand Binding to Proteins

C.R.BAGSHAW and D.A.HARRIS

1. INTRODUCTION

Biochemistry is to a great extent concerned with the interactions between proteins and other molecules. The latter may be other proteins (as in the interaction between actin and myosin), nucleic acids (as in ribosome assembly or repressor-gene interactions) or small molecules (as in enzyme-substrate or hormone-receptor interactions).

With few exceptions, specific interactions of proteins with other molecules involve a binding step as the prime event. The first stage in the characterisation of such molecular recognition requires quantitation of:

(i) stoichiometry of binding, i.e., the number of binding sites on the protein for the interacting species;

(ii) affinity of the binding sites for the ligand concerned.

Spectrophotometric methods are well suited to quantitative measurements of this type. Furthermore, such methods may provide other information about the protein-ligand complex, viz.

(i) kinetics of formation and/or breakdown;

(ii) structural changes in the protein following binding [e.g., binding of an activated receptor to a second protein (1)];

(iii) limited information on the nature of the binding site – its polarity, pK of ionisable groups, nearby residues, etc. The interpretation of these types of data, however, is often equivocal and needs to be supplemented by data from other methods, e.g., n.m.r., X-ray diffraction, chemical modification.

This chapter will deal with the application of standard laboratory spectrophotometers and spectrofluorimeters to study protein-protein and protein-ligand interactions. The stoichiometry of binding, the affinity of the binding sites for the ligand and the kinetics of formation/breakdown are covered in depth in this chapter. The detailed analysis of the structural changes in the protein following binding, and of the nature of the binding site is not discussed here.

2. APPLICABILITY OF PHOTOMETRIC METHODS

2.1 Advantages and Disadvantages

The measurement of ligand binding by spectrophotometry requires that the optical properties of the protein-ligand complex differ from those of the free ligand and free protein. Either the protein or ligand spectrum may change. A classical example is the measurement of oxygen binding to haemoglobin, using A_{577}, where the absorbance of the haem group changes on oxygenation.

A great strength of spectrophotometric methods is that the results may be obtained instantly and the progress of an experiment modified accordingly. This can be contrasted with equilibrium dialysis, where (a) the concentration range to be used must be estimated in advance and (b) the results are obtained hours or days later, during which time the protein may have denatured or the ligand decomposed. Another advantage is that the ligand and protein are allowed to equilibrate and measured in one vessel, so that membrane permeability problems, or rapid separations of the free and bound ligand, are avoided.

The main disadvantage of the use of spectroscopy to measure binding is that the optical signal is used empirically, i.e., the operator has no prior knowledge of how large an optical change (in protein or ligand) to expect on binding. Nor, if multiple binding sites are present, can it be assumed that the spectral change on binding a second or later ligand molecules will be identical to that observed with the first (see *Figure 9*).

Thus it may be difficult to establish the stoichiometry of protein-ligand interaction by spectroscopic methods alone. Often an initial calibration of the spectroscopic technique with direct (radioactive tracer) methods is useful; once the calibration is performed for an equilibrium study, for example, rates can be measured using spectroscopic methods only. Other approaches, however, may be used if such calibration is not possible (see Section 4.3).

2.2 **Choice of Technique**

Usually the technique to be used must be determined empirically for each system under study. However, some general principles can be given.

(i) Fluorescence methods are potentially several orders of magnitude more sensitive than absorption methods in absolute terms (see Chapter 3). Thus, fluorescence methods require less experimental material than absorption methods and may cover a wider concentration range.

(ii) Protein or ligand fluorescence is more sensitive than is absorption to the environment. Thus, the signal change (relative sensitivity) is likely to be larger on protein-ligand interaction if a fluorophore is observed. For example, ATP binding to myosin increases the absorbance of the protein by 2%, but its fluorescent emission by 20% (2,3).

Fluorescence changes of this magnitude or larger [NADH binding to lactate dehydrogenase gives a 3- to 5-fold enhancement in NADH fluorescence (see Section 4.2)], commonly result from transfer of a ligand into a less polar environment (water → protein interior) and are often accompanied by a perceptible blue shift in fluorescence emission. As noted in Chapter 3, however, fluorescence is also affected by other environmental factors (temperature, ionic strength, etc.) and careful control of such parameters is essential. In addition, at increasing ligand concentrations, a ligand may absorb light which would otherwise excite a protein fluorophore (or *vice versa*), and thus fluorescence will fall for this reason. This is known as an *inner filter* effect (see Chapter 1). Methods to correct for this effect are described later (Section 4.3).

In general, changes in absorption on binding (as a result of changes in polarity, etc.) are rather small and of limited use in spectrophotometric titrations. Fluorescence, despite its sensitivity to interference, must be the method of choice. However, there are a number of notable exceptions. If the chromophore can ionise, a shift in pK of the ionisable group on binding may lead to a large change in absorption. The ATP analogue, thioITP, has this property, and has proved useful in the study of myosin-nucleotide interactions (4) (see also Chapter 6).

(iii) In addition to absorption and fluorescence signals, light scattering (turbidity) may change on protein-ligand/protein-protein interaction. This signal is particularly useful when the reaction studied involves a change in the degree of molecular aggregation (actin-myosin, tubulin polymerisation, refs. 5,6). Turbidity may be monitored using a spectrophotometer or, more sensitively, monitoring 90° light scattering with a fluorimeter ($\lambda_{ex} = \lambda_{emit}$). Conversely, turbidity may cause artifacts – changes in absorbance/fluorescence emission may be observed or apparently enhanced by changes in turbidity due to protein aggregation, changes in membrane shape, etc. In this case, measurement at a reference wavelength in addition to the measuring wavelength may be needed (see Chapters 1, 2 and 5).

2.3 Chromophores and Fluorophores

2.3.1 *Protein*

All proteins contain natural chromophores absorbing in the near ultraviolet (see Chapter 3), notably the aromatic residues phenylalanine, tyrosine and tryptophan (λ_{max} = 260–290 nm). These chromophores are also responsible for protein fluorescence. In most proteins tryptophan fluorescence dominates (λ_{emit} = 340 nm), light absorbed by other aromatic residues being transferred to tryptophan non-radiatively. If tryptophan is low, or absent (e.g., in troponin C, F_1-ATPase), tyrosine fluorescence (λ_{emit} = 305 nm) may be observed.

Absorbance and/or fluorescence of these residues may change on binding protein to ligand. As noted above, absorbance changes are typically small but may be observable – particularly if tyrosine is transferred from an apolar to a polar environment (interior to surface of a protein), where its ionisation state may change (Section 2.2). Such effects may be amplified, as in solvent perturbation spectroscopy (7). Protein fluorescence changes are commonly used to measure protein-ligand interactions. Tryptophan fluorescence is often perturbed on ligand binding, for example in the binding of oligosaccharides to lysozyme (8).

Other chromophores, such as haem or flavin, are intrinsic to some proteins and provide useful spectroscopic 'handles' for studying protein-ligand interaction (*Table 1*).

2.3.2 *Ligand*

In principle, the large aromatic rings of the nucleic acid bases – both in nucleic acids and in their monomeric forms such as ATP – should be useful as strongly absorbing and fluorescent chromophores. Since a large number of important protein-ligand interactions involve the binding of nucleic acids or adenosine derivatives, for example, to proteins, one might expect a plethora of studies using spectroscopy and fluorimetry to study these interactions.

In fact, the absorbtion of these bases changes very little on interaction with proteins, and their intrinsic fluorescence is very low except in the case of a few unusual bases in tRNA (9). Thus, the spectral properties of purines and pyrimidines are of little use,

Table 1. Chromophores and Fluorophores.

	Absorption maximum (nm)	*Emission maximum (nm)*
Naturally occurring		
Adenosine	259	–
Tryptophan	280	340
Tyrosine	275	310
Phenylalanine (weak)	260	280
NADH	340	460
Flavin	450	520
Probes and analogues		
Thioinosine	340	–
N-Ethenoadenosine	300	415
Formycin	300	350
Dansyl derivatives	320	500
ANS	370	480
AEDANS derivatives	350	480
Fluorescein derivatives	480	530

in general, for monitoring binding of proteins.

The reduced pyridine nucleotides NADH and NADPH, however, are particularly useful. Their high intrinsic fluorescence has been mentioned previously (Chapter 3). Here we emphasise the sensitivity of this fluorescence to its environment; typically the fluorescence of NADH is enhanced some 3 – 10 times on binding to proteins. The binding of NAD(P)H to a variety of dehydrogenases has been studied fluorimetrically and an example is given in Section 4.3.

The enhancement of fluorescence when ligands bind to proteins, as noted above, is not usually a predictable property. However, it is generally true that a ligand becomes less mobile on binding to a protein as the ligand takes on the motional characteristics of the protein. Thus, the mobility of a ligand, as measured by fluorescence anisotropy, is a generally useful parameter in studying protein-ligand interactions. Relative changes are often large – the anisotropy of an ATP analogue on binding to F_1-ATPase rises from 0.06 to 0.24 (10). Antibody-antigen and drug-receptor interactions are routinely measured in clinical laboratories using such techniques (11). An apparatus for measuring fluorescence polarisation/anisotropy is shown in Chapter 1.

2.3.3 *Extrinsic Probes*

We see above that some natural ligands possess useful fluorescence/absorption properties for measuring binding and, in addition, all proteins contain potentially useful chromophores. However, the natural ligand may not be suitably chromophoric, and the observable changes in the protein too small for accurate measurement. In this case, extrinsic chromophores ('probes') may be introduced into the system.

Ligand analogues may be used as probes. A variety of fluorescent adenosine analogues [formycin, N-ethenoadenosine, 3′-O naphthoyl adenosine, etc. (*Figure 1*)] have been described, for example. The binding characteristics of the analogue itself can be deter-

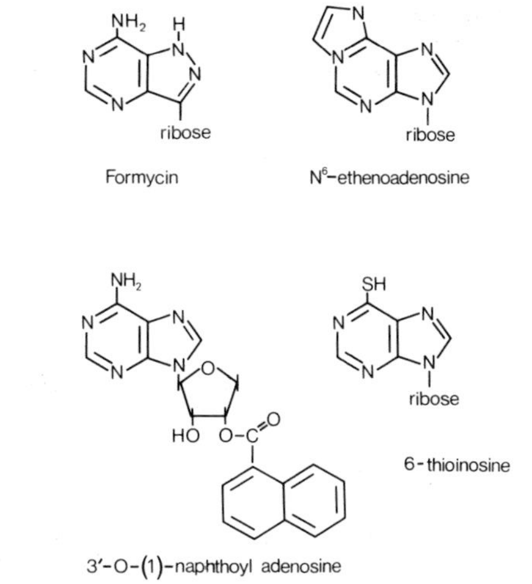

Formycin

N^6-ethenoadenosine

3′-O-(1)-naphthoyl adenosine

6-thioinosine

Figure 1. Chromophoric analogues of adenosine; non-covalent probes for nucleotide binding proteins.

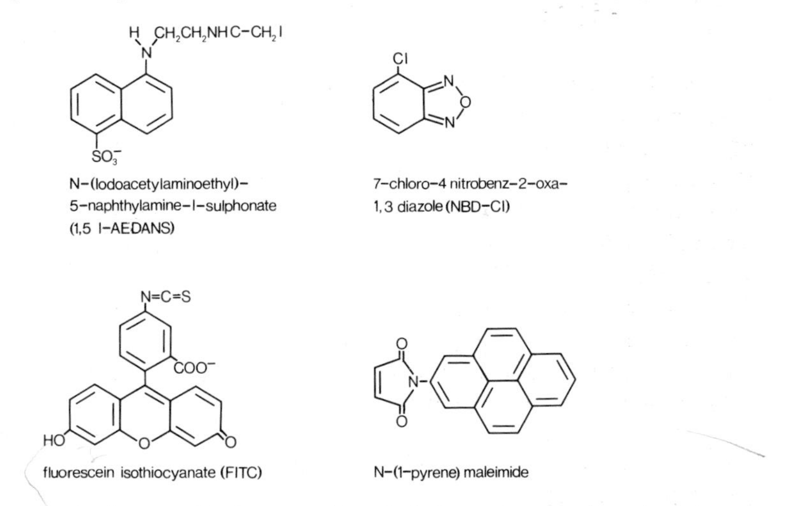

Figure 2. Covalent probes for fluorescent labelling of proteins; suitable for monitoring conformational changes in proteins.

mined spectroscopically, and this may yield information about the number of nucleotide binding sites on the protein and any interaction between them. Perhaps more importantly, binding characteristics of the natural ligand can then be studied by competition with the analogue, the second ligand displacing the chromophore and thus decreasing the signal (Section 4.2).

It may also be possible to label the protein covalently with a chromophoric 'reporter' group. Many reagents are available which react with amino or thiol groups on the protein, and allow the introduction of fluorophores, e.g., dansyl, fluorescein or pyrene groups (*Figure 2*), which may then respond detectably to changes in protein structure. These groups report on such changes only in molecules that are labelled. Thus, for an interpretable study of binding using these methods, it is necessary to establish:

(i) that labelling the protein does not affect its binding activity;
(ii) that the change observed is proportional to the amount of bound ligand.

Particular care is required when the labelling stoichiometry is low (<0.5 mol/mol) because although the optical signal from the derivatised protein may be adequate for spectroscopic studies, its altered biological properties may be masked by the unlabelled molecules. Nonetheless, useful information on the relationship between binding and conformational changes (in developing allosteric models) has been obtained by a combination of the above two methods (12,13).

Alternatively, non-specific ligands may be useful to label proteins non-covalently. Typically, the interaction of the protein with its true ligand may displace the first. 1,8-Anilino naphthalene sulphonate (ANS) is widely used in this context (see Chapter 5). This molecule is amphipathic and binds to many proteins and protein assemblies such as membranes. In aqueous solution its fluorescence is low, but this is enhanced dramatically when ANS is bound to a hydrophobic surface of a protein. This complex may respond to protein-ligand interactions [calmodulin-Ca^+ system (14)], or to protein-protein interactions [myosin light chain/heavy chain system (15)].

3. PRACTICAL CONSIDERATIONS

In its simplest form, a spectrophotometric titration involves the measurement of absorbance, or fluorescence, of a solution of protein upon successive additions of ligand. Typically, measurement occurs at a single wavelength and continues until no further change is observed (apart from change due to free ligand), when the enzyme is saturated with ligand (*Figures 3a* and *b*). In the case of the displacement of a probe/chromophore ligand by a natural ligand, the natural ligand is added to a solution of protein and probe at non-saturating probe concentrations (i.e., [probe] $\cong K_d > e_o$, see Section 5.1 and *Figure 3c*).

For optimal results (high signal/noise ratio), the following points should be considered.

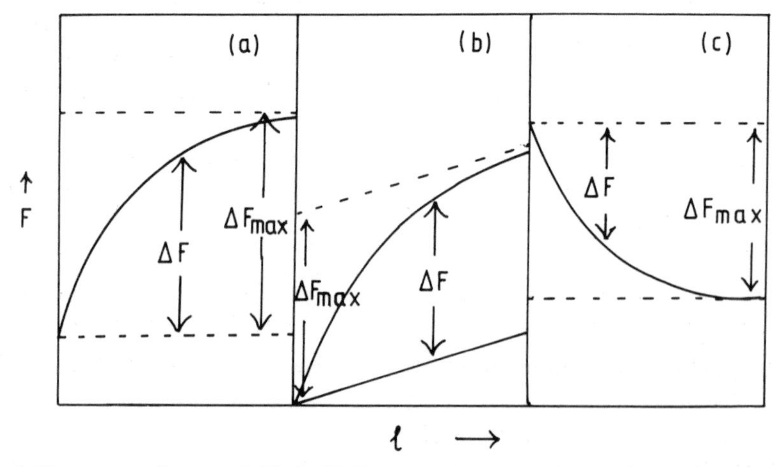

Figure 3. Fluorescence changes suitable for binding measurements. (**a**) Increase in protein fluorescence on ligand binding (cf. *Figure 9*); (**b**) Enhancement of ligand fluorescence on binding to protein (cf. *Figure 6*); (**c**) Competition of bound, chromophoric ligand with added, non-chromophoric ligand.

3.1 Lamp Stability

Lamp stability is a major instrumental limitation. Wherever possible, filament lamps should be used in preference to arc lamps. For example, when measuring absorbance in the region 325–350 nm where either a deuterium or tungsten-halogen light gives a suitable emission, the tungsten (filament) lamp is preferred.

Some compensation for lamp instability can be achieved by splitting off a fraction of the incident beam before the sample, and electronically comparing this beam with the transmitted/emitted beam. In spectrophotometers this is known as 'double beam' (split beam) operation. In fluorimeters, it is known as 'ratio' operation. In the latter case, only about 2% of the light is diverted from the sample as compared with 50% in the case of an absorbtion spectrometer (see Chapter 1).

Recent spectrophotometers (e.g., LKB Ultrospec 4050) and fluorimeters (Perkin Elmer LS5) use pulsed light sources which are internally calibrated and thus should not be so subject to lamp instabilities. The sample is also subject to less total irradiation, which may be an advantage if it is prone to photobleaching (see Section 3.3).

Lamp instability gives random fluctuations in the reading obtained.

3.2 Temperature Control

Temperature control ($\pm 0.1\,°C$) of solutions is important, especially in the case of fluorescence measurements, where both the signal itself and the equilibrium position of the binding are temperature dependent. Solutions should reach thermal equilibration before measurements are made or a regular drift in signal will be observed.

Plastic cuvettes are poor conductors and require longer to reach thermal equilibrium than glass or quartz. If problems of temperature control are suspected, it is advisable to check the temperature of the solution in the cuvette rather than to rely on the temperature of the water bath.

3.3 Photodecomposition

Some absorbing compounds undergo chemical changes on illumination and this will result in a gradual change in absorbance or fluorescence in the cuvette during the measurement. This is more often a problem with fluorescent compounds which undergo photobleaching, but some other compounds (e.g., NBD-Cl) darken perceptibly when exposed to light.

In the short term, such effects can be minimised by stirring the cuvette to remove bleached material from the light path. In the longer term, the only solution is to allow less light to fall on the sample by:

(i) reducing the slit width;
(ii) working in the dark and closing the photometer shutter between measurements;
(iii) using an instrument with a lower beam intensity (see Section 3.1).

Both temperature disequilibrium and photodecomposition give a regular drift in the reading obtained.

3.4 Turbidity

Turbidity in the solution leads to increased light scattering, which is deleterious to absorbance and (especially) fluorescence measurements. There are two possibilities of interference.

(i) Large particles of material may scatter light and sink down into the light beam, leading to large random noise peaks, especially in fluorescence measurements. Solutions should be filtered prior to use through a 0.45 μm filter (e.g., Gelman Sciences Acro LC3) or clarified by ultracentrifugation.

(ii) The protein itself may denature/aggregate over the course of the experiment. This can normally be detected by eye as an increasing cloudiness and is seen in the spectrophotometer as a regular upward drift in reading. The stability of the protein over the period of the titration should be checked, and visual inspection of the cuvette contents before and after the experiment should be routine.

A common cause of denaturation is the protein meeting a very high local concentration of ligand or its solvent, and it may be necessary to stir the cuvette continuously and/or add ligand from a more dilute stock solution. Alternatively, the titration can take the form of protein *versus* (protein + ligand), where the latter is made up to contain double the final ligand concentration required. Initially, the protein alone is placed in the cuvette (e.g., 2 ml) and aliquots (say, 0.1 ml) of the (protein + ligand) solution are added to give a total volume of 4 ml. In this way, the ligand concentration is progressively increased without incurring any dilution of protein or local high concentration of reactants.

4. MEASUREMENT OF PROTEIN-LIGAND EQUILIBRIA

4.1 **Principle**

First a wavelength (or wavelength pair for fluorescence measurement) is selected at which the change in (protein and ligand) spectrum is maximal on combination. The protein is then titrated with ligand while the optical change is followed at this wavelength. Procedures for calculating the dissociation constant K_d and the number of binding sites per protein molecule (n) are given below.

4.2 **Selection of Wavelength and Slitwidth**

4.2.1 *For Absorbance Changes*

Where absorption spectroscopy is employed, it is likely that the change in signal on interaction is relatively small (see Section 3). In this case, use of a double beam/split beam spectrophotometer (or computer storage of spectra) to produce a difference spectrum directly is almost mandatory (see Chapters 2 and 5). Use of such an instrument is described.

(i) Measure the spectrum of the buffer alone using pure water as a blank. The buffer absorbance should be low (ideally $<10\%$ of the sample absorbance) over the region of interest. This is rarely a problem if measurements are to be made in the visible region (and is obvious by eye if it is!), but may be a problem in the u.v. If the buffer does absorb significantly, use a different buffer.

(ii) Fill a tandem cell (e.g., Hellma no. 230 from Hellma, Westcliff-on-Sea, Essex, UK), or a pair of conventional cells mounted in line in the sample holder, separately with protein and ligand solutions at the highest required concentration. Place an identical tandem arrangement in the reference beam, and scan the spectrum at the highest gain setting, with the pen offset to the centre of the recorder chart.

Ideally this record should be a straight line. If the pen judders excessively, the absorbance of the solutions is probably too high and insufficient light is reaching the detector. The solutions should then be diluted. If, on the other hand, the pen shows systematic deviations, a difference between either the cuvettes or the solutions in the two beams is indicated. Differences in the cuvettes can be minimised by careful selection, but note that the difference may lie in the pathlength rather than the material of construction – only the latter is detected by 'matching' the cuvettes filled with water. Differences between the solutions are minimised by filling both sides of the system from single stock solutions of ligand or protein.

(iii) Once a satisfactory baseline has been recorded, mix the protein and ligand solutions in the sample beam and measure their spectra against the unmixed reference solutions. With this arrangement, although the sample and ligand are diluted twice, the effective path length for each component is doubled and any peaks or troughs in the difference spectrum indicate potentially useful wavelengths for monitoring complex formation. As a check, mix the contents of the reference cells and record the difference spectrum again; the baseline spectrum should be re-established.

(iv) Select a wavelength and slitwidth for further experimentation. The wavelength setting will correspond to the maximum peak (or trough) observed in the difference spectrum. The slitwidth should be as wide as possible without exceeding the natural bandwidth of the peak or trough under investigation, so as to maximise the amount of light reaching the detector (Chapter 2).

If the observed change is sufficiently large, subsequent measurements may be made using a single beam instrument at this fixed wavelength. However, double beam/split beam instruments are convenient, particularly when the concentration of the absorbing species (generally ligand in this case) is to be varied during the titration.

4.2.2 *For Fluorescence Changes*

For fluorescence measurements, two wavelengths and two slitwidths (one of each for excitation and emission beams) must be chosen. The situation is further complicated since double beam fluorimeters are not generally available and summation/subtraction of spectra must be performed manually (or, again, using computer storage of spectral data).

(i) Choose an excitation wavelength on the edge of an absorption peak of the species under investigation (NADH below). Although this reduces emission intensity, it allows higher concentrations of chromophore to be used without incurring unacceptable inner filter effects (see Section 2.2). If possible, the excitation slit is kept narrow (<5 nm) to reduce photodecomposition and to achieve sufficient resolution to avoid strong absorption bands [but see step (iii)].

(ii) Measure the emission spectra of (a) water and (b) buffer over the range of interest. Fluorescence should be low ($<10\%$ of the expected signal). If the water itself shows fluorescence, it should be pretreated with active charcoal (e.g., as

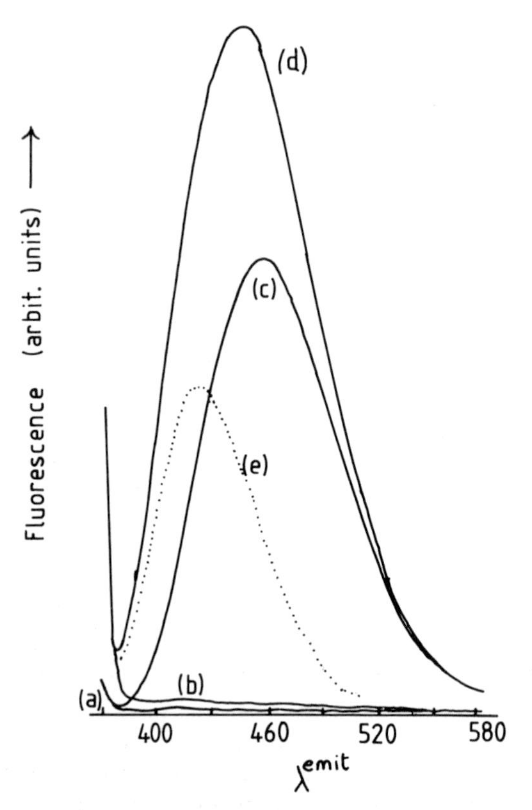

Figure 4. Fluorescence emission spectra of the components of the lactate dehydrogenase (LDH)/NADH system. 90° fluorescence was measured using a Perkin-Elmer MPF-2a spectrofluorimeter, with instrument settings λ_{ex} = 360 nm, excitation slit = 6 nm, emission slit = 16 nm, and λ_{emit} was varied from 380−520 nm. The cuvette contained 2 ml of: (**a**) 0.1 M potassium phosphate buffer pH 7 (phosphate buffer); (**b**) LDH from pig heart (0.46 mg/ml) dialysed into phosphate buffer; (**c**) NADH (25 μM) in phosphate buffer; (**d**) LDH (0.46 mg/ml) + NADH (25 μM) in phosphate buffer. Note the increase in light scattering (especially at lower wavelengths) in the presence of enzyme, and the enhancement and blue shift of NADH fluorescence on binding to the enzyme. Curve (**e**) represents the difference between curves **c** and **d**, viz. the fluorescence of bound NADH. From this, λ_{emit} was chosen to be 445 nm in subsequent measurements. All measurements were made at 20°C.

in the Millipore Super Q water purification system), prepared fresh, and kept in acid washed *glass*ware. Plastics contain fluorescent materials which leach into the water on standing. If the buffer is fluorescent, purification (active charcoal, recrystallisation) may be possible, but is generally easier to buy a purer reagent grade.

(iii) Measure the emission spectrum of the 'observed' species (NADH below) at the highest concentration to be used, and adjust the emission slit (5−20 nm) and sensitivity controls to establish the observed peak on a suitable scale for measurement. If the sensitivity is too low, the excitation slit may be widened, or λ_{ex} moved closer to the absorbance maximum [cf. step (i)].

(iv) Measure the emission spectrum of the other species (protein below) at the highest concentration to be used, and of the (protein + ligand) mixture. The difference between the spectrum of the mixture, and the sum of the two separate spectra

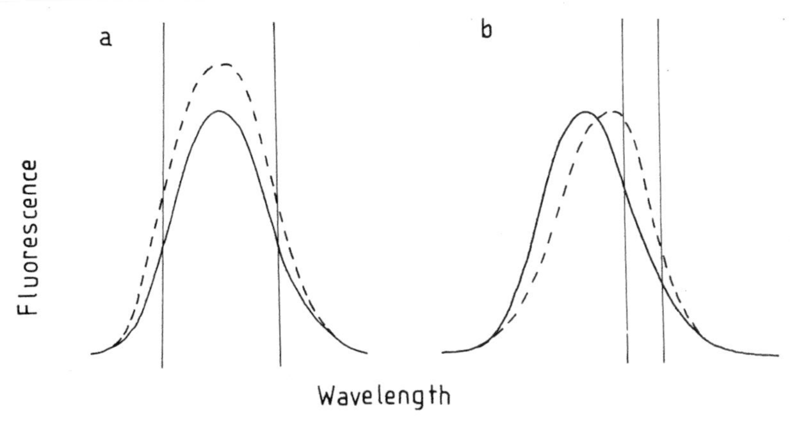

Figure 5. Optimisation of wavelength and slit width. The optimal wavelength and slit width for maximising the difference signal on ligand depends on the extent to which the amplitude is changed and the peak wavelength is shifted. Emission spectra in the absence (———) and presence (- - - - - -) of the interacting substance are first recorded using a narrow slit-width. In (**a**) interaction causes an increase in fluorescence with little change in peak wavelength. A wide slit can therefore be used to follow binding with the advantage of increased light output (i.e., better signal-to-noise). In (**b**), the shift in wavelength on interaction indicates that the maximum signal change will be achieved by setting the emission wavelength on the edge of the spectrum. A narrow slit must be selected. Note that in practice, the band pass of a monochromator is triangular rather than rectangular but the same principle applies. In the case of absorption spectroscopy, double-beam instruments allow the difference spectrum to be measured directly (see text), which aids selection of the optimal wavelength and slit width.

(after correcting for the buffer, if necessary) yields a difference spectrum from which the wavelength (λ_{emit}) of maximum change can be determined (*Figure 4*).

(v) If the perturbation on ligand binding involves a change in amplitude (enhancement, quenching of fluorescence), take λ_{emit} as the peak position on the emission spectrum and use a wide slitwidth ($\geq$ 10 nm). If, however, the peak is merely shifted, take λ_{emit} on the edge of the emission peak and use a narrow slit. *Figure 5* shows the difference between the two situations.

(vi) Ideally, the fluorescence of water, buffer, and the 'non-observed' species should be negligible at the wavelength pair under study. Calculation is thus considerably simplified. If however a background fluorescence, due to these species, is observed, a constant 'back-off' voltage may be applied to the fluorimeter output to subtract this signal from the varying signal, thus widening the usable range of the instrument.

4.2.3 *For Light Scattering Changes*

For measurements of light scattering, select a wavelength as low as convenient for maximum sensitivity (scattering being proportional to $1/\lambda^4$), but at a position clear of any absorption bands. For a system containing only proteins and/or nucleic acids, 325–360 nm is a convenient region.

4.3 **Titration of Lactate Dehydrogenase with NADH (16)**

As an example of a spectrophotometric measurement of binding, the use of NADH fluorescence to monitor its binding to L-lactate dehydrogenase (LDH) is described here.

It is assumed that suitable wavelengths and slitwidths for this system have been determined as above, and that our knowledge of the enzyme properties have allowed us to select a suitable buffer, temperature and concentration range compatible with enzyme stability and physiological function.

LDH catalyses the reaction

$$NAD^+ + \text{L-lactate} \rightleftharpoons NADH + \text{pyruvate}.$$

The enzyme binds NADH in the absence of pyruvate, and binding is associated with a variety of optical changes:

(i) a small decrease in NADH absorbance;
(ii) an enhancement, and a blue shift of NADH fluorescence;
(iii) a quenching of protein (tryptophan) fluorescence.

Any of these effects can be used to monitor NADH binding to this enzyme, and the principles, by extension, to measure the binding of any other chromophore to any other protein.

Since the enhancement of NADH fluorescence on binding is large (about 3-fold), this is used below to monitor binding.

4.3.1 *Method*

(i) Dialyse LDH (pig heart) against 0.1 M potassium phosphate pH 7.0 (phosphate buffer) prior to use.
(ii) Add 2 ml of 3.5 μM (0.5 mg/ml) LDH in phosphate buffer to a fluorimeter cuvette and allow to equilibrate at 20°C. Prepare a similar cell containing 2 ml of phosphate buffer alone.
(iii) Set λ_{ex} at 360 nm (edge of NADH absorption band), λ_{emit} at 445 nm (peak of NADH difference spectrum), as determined in Section 4.2.
(iv) Add 5 or 10 μl aliquots of NADH (500 μM) sequentially to each cuvette over the range 1.25 – 12.5 μM NADH (final concentration). After each addition, stir the cuvette and record F_{445}. The range can be extended to 125 μM NADH by using a second NADH solution of concentration 5 mM. (The total volume added here is kept below 0.1 ml, to minimise corrections for dilution – but see Section 3.4.)
(v) Plot the fluorescence after each addition of NADH to yield the two curves shown in *Figure 6*. After correction for dilution, the lower curve should be linear – if not, an inner filter effect is occurring (see Section 2.2) and the experiment should be repeated using λ_{ex} slightly higher (say, 365 nm). [Inner filter effects should be negligible if the absorbance at the excitation wavelength is less than 0.1. If inner filter effects cannot be avoided over the concentration range of interest, a semi-empirical method of correction is available (18). See *Figure 8* and legend.]
(vi) Calculate the difference between the two curves (*Figure 6c*). This shows the fluorescence enhancement at each NADH concentration (ΔF_c), which is proportional to bound [NADH].
(vii) Also required for the analysis below is ΔF_{max}, the enhancement at saturation. This is proportional to the concentration of total NADH binding sites. This can

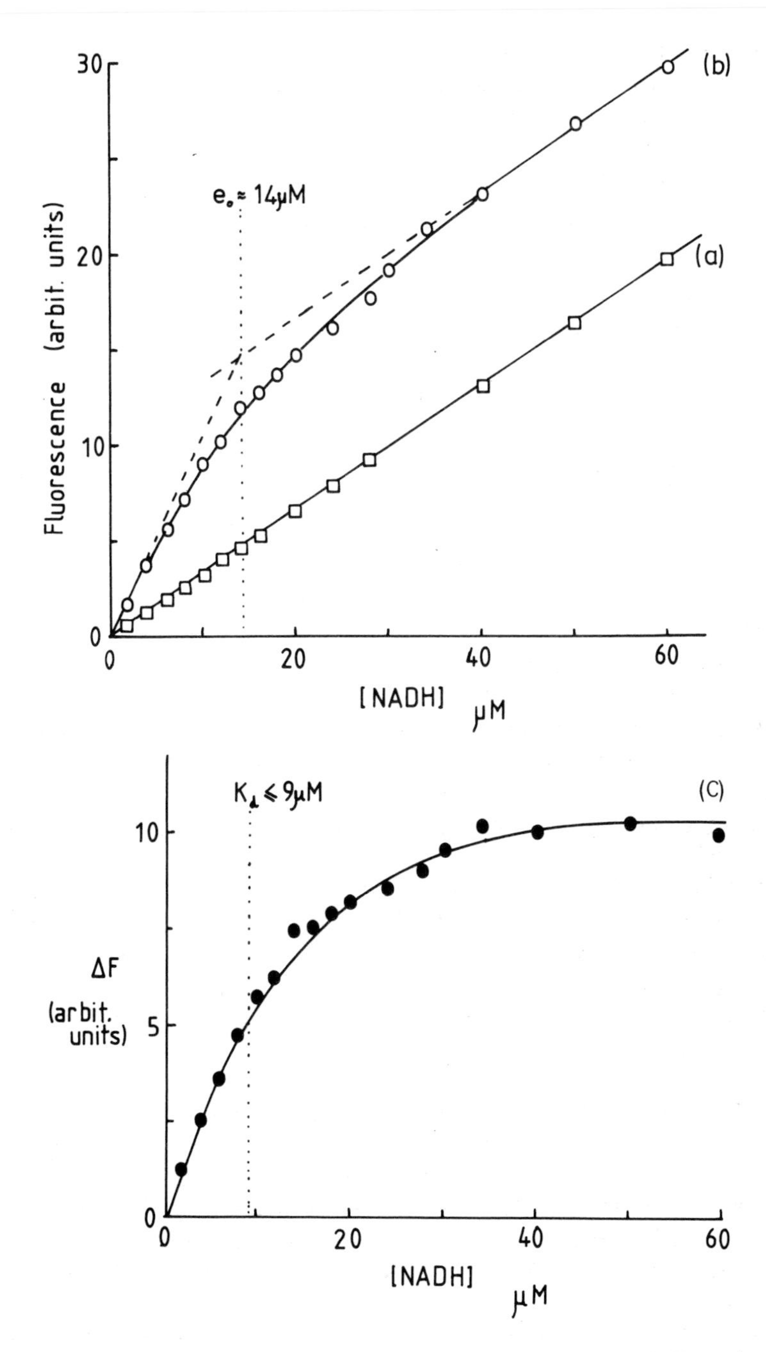

Figure 6. Fluorescence titration of LDH with NADH. Measurements were made as in *Figure 4*, except that λ_{emit} was fixed at 445 nm, and NADH levels were varied from 0–60 μM by addition from a concentrated stock solution. After each addition of NADH, the mixture was stirred and the (steady) value of fluorescence recorded. The overall volume change was less than 5%. **Curve a**: cuvette contained initially 2 ml phosphate buffer. **Curve b**: cuvette contained initially 2 ml 0.46 mg/ml LDH in phosphate buffer. **Curve c**: difference between **curves a** and **b** viz. increase in fluorescence due to bound NADH.

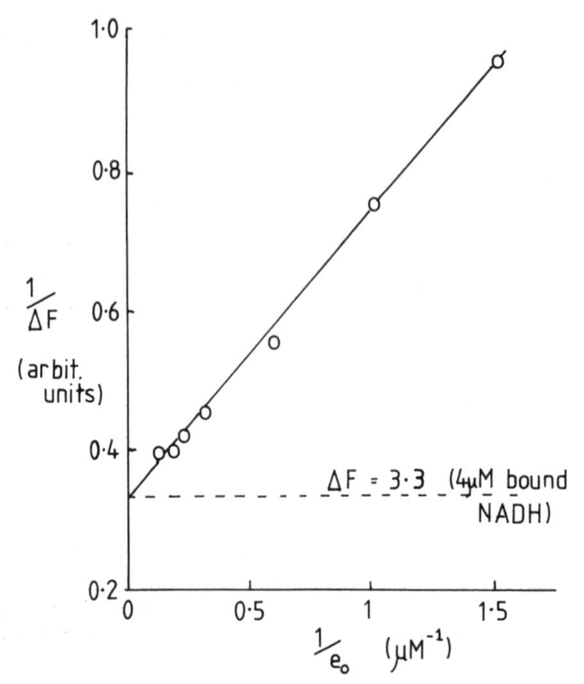

Figure 7. Estimation of the fluorescence of bound NADH. Measurements were made as in *Figure 6*, except that NADH concentration was set at 4 μM, and LDH concentration varied by addition from a stock solution of 9.2 mg/ml LDH in phosphate buffer. The points at high concentrations are corrected for dilution, the correction being less than 10% at all enzyme concentrations. Extrapolation to infinite enzyme concentration indicates an *increase* in NADH fluorescence on binding of 3.3 arbitrary units for a 4 μM solution, i.e., an *increase* in fluorescence of 0.82 arbitrary units per μM NADH bound.

be estimated from a plot of $1/\Delta F_c$ *versus* 1/[NADH], which approaches linearity at high [NADH] and crosses the ordinate at $1/\Delta F_{max}$.

(In general, it may be possible to estimate ΔF_{max} from the difference curve by eye if the unbound ligand gives no, or very little, fluorescent signal itself (see *Figure 3*), and can thus be raised to $>20\ K_d$ in concentration. Alternatively, an independent estimate of ΔF_{max} may be made as in Section 4.3.2.)

The aim here is to measure the equilibrium position between bound and free NADH. For most binding reactions in solution, equilibration takes place within 100 msec and so measurement can be immediate using manual techniques. However, binding reactions may be slow, particularly if protein-protein interactions are under study, and in this case it is essential that equilibrium is reached before measurement is recorded. Equilibrium is indicated by a reading unchanged in time.

4.3.2 *Modifications*

(i) *Independent determination of* ΔF_{max}. If ΔF_{max} cannot be accurately estimated from the above data (i.e., saturation requires an NADH concentration too high for the instrumental capability), it can be measured independently. In this case, NADH concentration in the cuvette is fixed, say at 4 μM, and the enzyme added in aliquots up to,

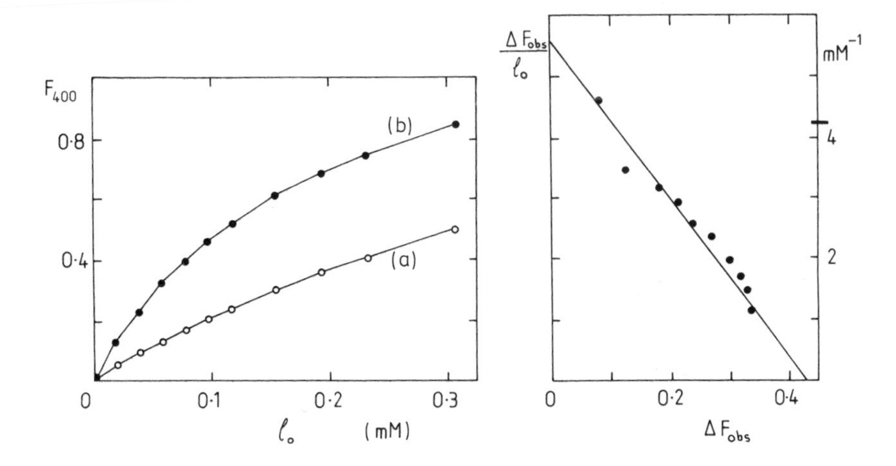

Figure 8. Correction of binding measurements for inner filter effect. Curve (**a**) shows the fluorescence observed when nicotinamide 1,6-N ethanoadenine dinucleotide phosphate (ϵ-NADP$^+$) is added to buffer (λ_{ex} = 340 nm, λ_{emit} = 400 nm). The increase in fluorescence is non-linear due to the inner filter effect (cf. *Figure 6*). Curve (**b**) shows the fluorescence observed when dihydrofolate reductase was also present. An enhancement of fluorescence on ligand binding is observed. Subtraction of curve (**a**) from curve (**b**) yields data (ΔF_{obs}). A plot of $\frac{\Delta F_{obs}}{l_o}$ versus ΔF_{obs} yields a line of slope $-1/K_d^{app}$ and intercept ΔF^*_{obs}, corresponding to the increase in fluorescence when all enzyme is saturated with ligand. (This treatment assumes that $l \simeq l_o$, the amount of bound ligand being negligible. While this is a fair approximation in this case, it is not in the data of *Figures 6* and *11*). From this line, K_d^{app} = 0.078 mM.
Due to the severity of the inner filter effect (about 30% attenuation at the highest ϵ-NADP$^+$ concentration), the curvature of the difference curve will be greater than that of the true binding curve, and K_d will be underestimated ($K_d^{app} < K_d$). To correct, we assume ΔF_{true} is proportional to l_o in the absence of enzyme. (i) If the inner filter effect is small at low l_o, $\Delta F_{true} \cong \Delta F_{obs}$, and a theoretical line can be drawn through the initial points on curve (**a**). At higher values of l_o, a correction factor ($\Delta F_{obs}/\Delta F_{true}$) can be calculated by comparing the extrapolated theoretical line with the experimental curve at any added ligand concentration l_o. The upper curve can then be corrected by dividing ΔF_{obs} by this correction factor at the appropriate l_o (i.e., *proportional* decrease is taken as the same for upper and lower curves). (ii) In the more general case, curve (**a**) is fitted to the polynomial

$$\Delta F_{obs} = \Delta F_{true} \frac{e^{-adl_o} - e^{-al_o}}{al_o^{(1-d)}}$$

where ΔF_{true} is again taken to be proportional to l_o ($\Delta F_{true} = kl_o$), and a and d are constants, (a related to the extinction coefficient for ligand, and d to the fraction of cell width from which measurements are taken). Once a and d are determined from curve (**a**), the correction factor $\Delta F_{obs}/\Delta F_{true}$ can be calculated for any value of l_o and applied to curve (**b**) as above. Using this correction, K_d from the above data was calculated as K_d = 0.17 mM (cf. above). [Data and correction procedure from Birdsall *et al.* (18)].

say, 20 μM, when nearly all the NADH will be bound. A plot of $1/\Delta F$ against $1/e_0$ is made, and extrapolated to $1/e_0 = 0$ (*Figure 7*), when $\Delta F = \Delta F_{max}$ for 4 μM NADH.

(ii) *Independent determination of number of binding sites.* Inclusion of 0.6 mM oxamate (a pyruvate analogue) in the enzyme solution decreases K_d for NADH by several orders of magnitude. Under the conditions above, therefore, inclusion of oxamate will cause nearly all the NADH added to be bound. This may be useful in the determination of the number of binding sites per mol enzyme, n (see Section 5.1 and *Figure 10*).

(iii) *Monitoring protein fluorescence.* NADH binding can also be followed by measuring the quenching of tryptophan fluorescence of the protein. The experimental set up

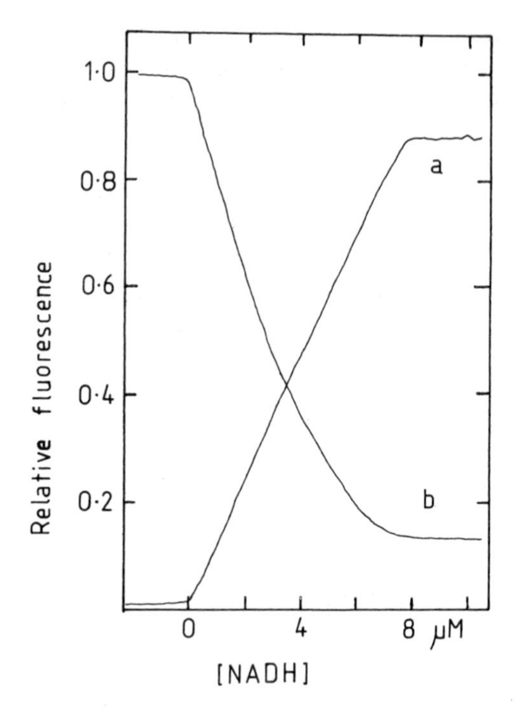

Figure 9. Binding of NADH to LDH; comparison of NADH and protein fluorescence. NADH was added continuously from motor driven syringes into two cuvettes containing 10 mM oxamate solution and, in the sample cuvette, 2 μM LDH. Fluorescence was monitored using a differential fluorimeter (16), the cuvette without enzyme being used for reference. Curve (**a**) shows difference in NADH fluorescence, indicating that K_d for NADH is very low in the presence of oxamate ($K_d \ll 2$ μM), and that the binding stoichiometry is 4 mol NADH/mol protein. Curve (**b**) shows the quenching of the protein fluorescence in an analogous experiment, except that the reference cuvette was set up to monitor the transmission of the exciting light so as to correct for the inner filter effect (see also *Figure 8*). Curve (**b**) indicates a value of K_d ~0.2 μM with a stoichiometry of about 2 mol NADH/mol protein. However, these values are distorted because the quenching of protein fluorescence is not linearly dependent on NADH binding. The first NADH bound quenches the fluorescence of the binding subunit and that of its neighbours – thus measurement of quenching leads to an underestimate of binding stoichiometry. In the later stages of the titration, NADH binding has less effect on the relative fluorescence, because the neighbouring subunits are already quenched by the bound NADH. Thus the curvature that results can give rise to an erroneous estimate of K_d. [A full analysis of the problem is given in ref. (16), from where these data are derived.]

is as described above except that fluorescence is measured using λ_{ex} = 295 nm and λ_{emit} = 340 nm. This method has the advantage that NADH alone gives no signal and subtraction is not necessary (cf. *Figure 6*, above). However, this method has a general and a specific complication.

The general complication (occurs in most systems) is that inner filter effects are likely to be significant, i.e., the aromatic ligand absorbs light which would otherwise excite the protein. To control for this, tryptophan or *p*-aminobenzoyl glutamate (17), at concentrations such as $A_{295} \cong A_{295}$ of the protein, should be included in the blank. The calculation is then carried out as in *Figure 8*.

The specific complication of the LDH system is that the quenching of protein fluorescence on NADH addition is not linearly related to NADH binding. Binding the first molecule of NADH apparently quenches tryptophan not only in the occupied but also in the neighbouring subunits. This was shown by Holbrook (16), who monitored NADH fluorescence enhancement and protein quenching in the same experiment. NADH fluorescence enhancement is linearly related to binding, protein fluorescence quenching is not (*Figure 9*).

This complication of the LDH system emphasises a more general point made in Section 2.1 above, viz. that it cannot be assumed, without some initial calibration, that each molecule of ligand binding to a multisite enzyme will give the same optical change even if the sites are functionally identical (independent).

(iv) *Determination of K_d for competing ligands.* If NADH concentration is poised such that the enzyme is about half saturated ($\Delta F = 1/2\Delta F_{max}$), addition of a competing ligand (say, NAD^+) will displace NADH and ΔF will decrease (if $K_d \sim [NADH] \gg e_o$, *Figure 3*).

By a standard analysis

$$\Delta F_c = \frac{\Delta F_{max}}{1 + \frac{K_d}{l}\left(1 + \frac{i}{K_i}\right)} \qquad \text{Equation 1}$$

(where l is the concentration of ligand producing the signal, i the concentration of competing ligand and K_i its dissociation constant).

This is a plot of $\Delta F_{max}/\Delta F_c$ against i yields a straight line, with intercept $[1 + (K_d/l)]$ on the ordinate and slope $(K_d/l) \times (1/K_i)$. The dissociation constant of the competing ligand, K_i can thus be determined without independent knowledge of K_d. This procedure is useful if, for example, a fluorescent analogue of the substrate is available whose properties are of limited intrinsic interest except as a marker for binding of the true substrate.

5. TREATMENT OF BINDING DATA

The following symbols are used in the discussion below.

K_d = dissociation constant of the protein-ligand complex
n = number of binding sites per protein molecule
e_o = total concentration of binding sites for ligand
e = concentration of free binding sites at equilibrium
l_o, l = total, free concentration of added ligand at equilibrium
b = concentration of bound ligand
α = fraction of enzyme bound to ligand = $\frac{e_o - e}{e_o}$

By definition $K_d = \frac{e.l}{b}$ Equation 2

5.1 Rough Determination of K_d and n

Under limiting conditions, K_d or n can be estimated from the untransformed data obtained above. If binding is relatively weak ($e_o < 0.1\ K_d$), very little ligand is bound in the initial stages of titration ($l_o \sim e_o$), and l, the free ligand concentration $\sim l_o$.

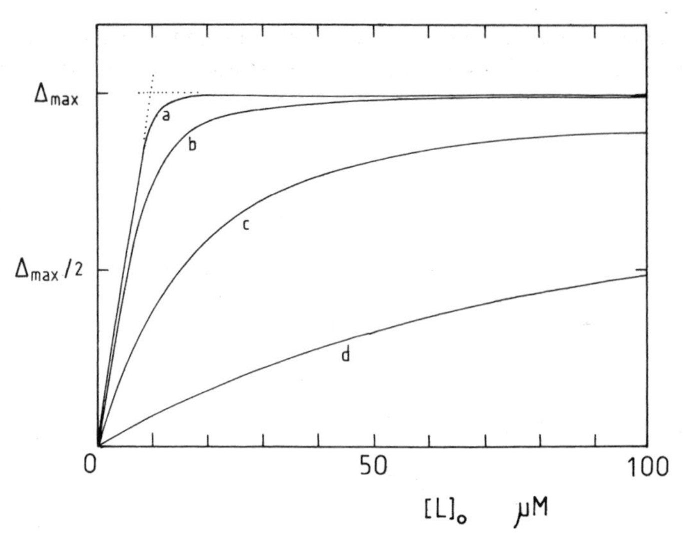

Figure 10. Theoretical spectroscopic titration curves. The change in signal, Δ, was calculated from Equation 2, assuming a linear dependence of Δ on b. The total protein concentration was taken as 10 μM (e_o), and the dependence of Δ on l_o is shown for values of K_d = (a) 0.1 μM; (b) 1 μM; (c) 10 μM and (d) 100 μM. When $K_d \ll e_o$, stoichiometry of binding can be estimated from the break point – here at l_o = 10 μM, thus indicating a 1:1 stoichiometry (**Curve a**). When $K_d \gg e_o$, K_d can be estimated from the value of l_o at half maximal signal change – here $K_d \cong$ 105 μM (**curve d**). Only rough estimates of K_d and stoichiometry can be obtained from curves (**b**) and (**c**) by inspection (cf. *Figure 6*). More accurate values can be obtained by alternative treatments (cf. *Figure 11*) or by a reiterative fit.

Thus, $K_d = l_{1/2}$, the ligand concentration that gives half-maximal reponse. (At lower K_d values, $l_o > l$, and K_d is overestimated by this method.)

Conversely, if binding is strong ($e_o > 10\ K_d$), almost all ligand added in the early stages of the titration is bound, and $l \ll l_o$. In this case n can be measured from the intersection point of the initial rise and the plateau of the response. These two cases are given in *Figure 10*.

In principle, it should be possible to arrange conditions so that n and K_d may be measured, in the same instrument, by separate titrations. This may be achieved by varying e_o or by varying K_d – the latter by changing the buffer composition or temperature. K_d for NADH binding to LDH can be determined in the absence of oxamate, and n in its presence, for example (Section 4.3). However, these options are rarely available in practice since signal to noise ratios limit the amount of protein and/or ligand used within fairly narrow limits.

5.2 Graphical Determination of K_d and n

Conditions are chosen so that $e_o \sim K_d$ – a significant amount, but not all of the ligand added is bound. The data are conveniently plotted in linear form using the following analysis (20)[1]. Linearising the data makes it simpler to detect deviations from expected

[1]A more familiar linearisation, due to Scatchard, gives

$$\frac{b}{l} = -K_d(b) + e_o$$

However, since neither l nor b is known directly from optical titrations the above method requires less manipulation of the available data. The data of *Figures 6* and *7*, recalculated in this form are plotted in *Figure 11b*.

behaviour, either due to non-proportionality between binding and optical response or to non-equivalence in binding sites present. To establish which of these two possibilities occurs, direct calibration of binding is necessary.

We have

$$K_d = \frac{e.l}{b} \qquad \text{Equation 2}$$

If a fraction α of the protein is bound to ligand at equilibrium, then:

$$e = e_o(1-\alpha)$$
$$b = \alpha e_o$$
$$l = l_o - b = l_o - \alpha e_o$$

Substituting in Equation 2

$$K_d = \frac{e_o(1-\alpha)\,(l_o - \alpha e_o)}{\alpha e_o} \qquad \text{Equation 3}$$

$$K_d = \frac{1-\alpha}{\alpha}\,(l_o - \alpha e_o)$$

or

$$\frac{l_o}{\alpha} = \frac{K_d}{(1-\alpha)} + e_o \qquad \text{Equation 4}$$

The variables in this equation, in a titration of enzyme with varying amounts of ligand, are α and l_o. l_o is simply the concentration of added ligand. α, the fraction of enzyme sites occupied by ligand, is given by $\alpha = {}^{\Delta}/_{\Delta_{max}}$ where Δ is the optical change measured. This determination of α (Section 4.3) at different l_o values allows a plot of

$$\frac{l_o}{\alpha} \textit{ versus } \frac{1}{1-\alpha}$$

from which K_d (slope) and e_o (intercept) can be measured. e_o represents the concentration of ligand binding sites on the enzyme present. Comparison of this value with molarity of the enzyme yields n, the number of binding sites per enzyme molecule (*Figure 11a*).

If deviations from linearity occur and these can be correlated with binding effects (not necessarily straightforward), a concave downwards curve is indicative of positive cooperativity in binding (K_d increases as saturation decreases), while a concave upwards plot indicates negative cooperativity or, more probably, a mixture of binding sites (the weaker binding sites filling last – K_d^{app} rises as saturation increases).

6. TIME-DEPENDENT REACTIONS

Spectroscopy is well suited to the measurement of time-dependent reactions because it produces electrical signals which can be monitored continuously. Spectroscopy is widely used to monitor enzyme activity in the steady-state as outlined in Chapter 3, and its application to rapid processes, in pre-steady-state analysis, is described in Chapter 6. The majority of enzymes bind ligands and turn over their substrates on a time scale

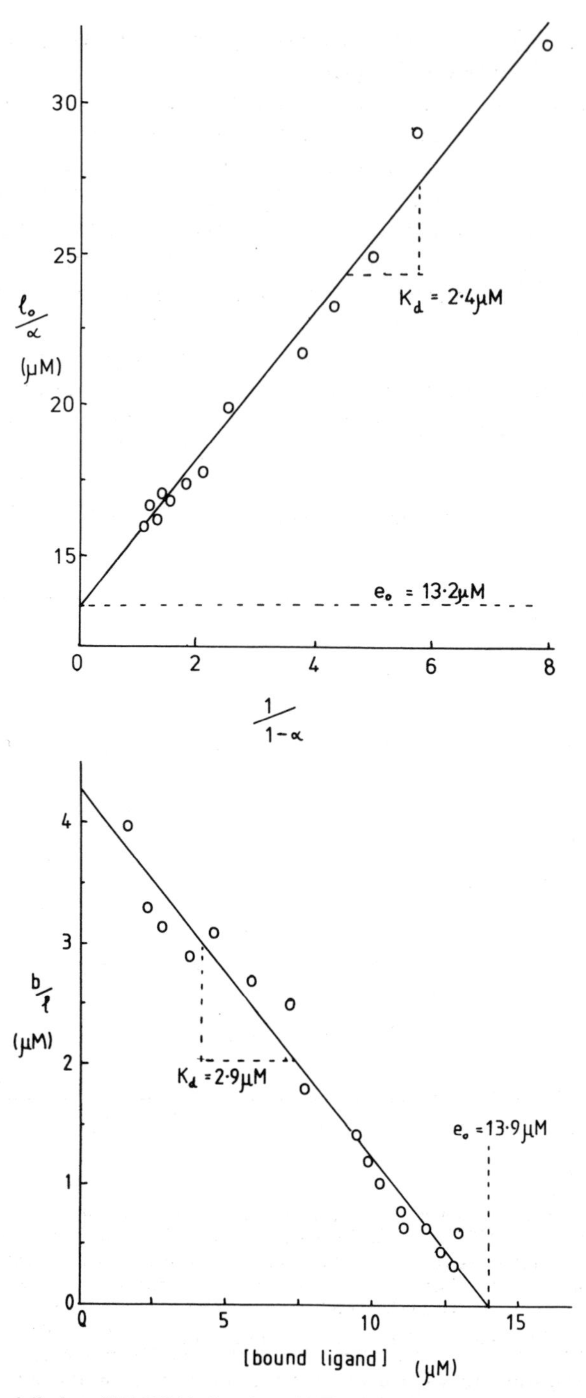

Figure 11. Graphical display of NADH binding data. (**a**) The data of *Figure 6* are plotted according to Equation 4, with $\alpha = \Delta F/\Delta F_{max}$, and l_o = concentration of NADH added. (**b**) The data of *Figure 6* are plotted according to the Scatchard equation $b/l = -b/K_d + e_o$. b is calculated from ΔF, assuming the value of ΔF for bound NADH to be 0.82 arbitrary units per μM (*Figure 7*). l is calculated from $l = l_o - b$.

of less than 1 sec. To study such reactions requires rapid mixing devices and rapid recording spectrophotometers (Chapter 6).

However, some biochemical processes may involve interactions with half-times of the order of seconds or minutes. These include conformational changes in some proteins, protein-protein aggregation and dissociation of tightly bound ligands. For example, certain repressors bind to DNA with a K_d of less than 1 nM. Since the association rate must be at, or slower than, the diffusion controlled limit (10^8 M^{-1} sec^{-1}), dissociation must take several seconds or minutes to occur. In this case the process (if associated with a suitable spectroscopic change) can be followed in a conventional spectrophotometer.

In the case of enzyme kinetics, there are certain advantages in studying the enzyme intermediates directly rather than by inference from steady-state analysis. Conventional spectrophotometry may be employed to study single turnovers of enzyme activity when turnover is suppressed by a natural or synthetic inhibitor. For example, after covalent modification of an enzyme, steady-state kinetics may reveal that its activity is reduced by 99%. This leaves the question as to whether the 1% residual activity represents the

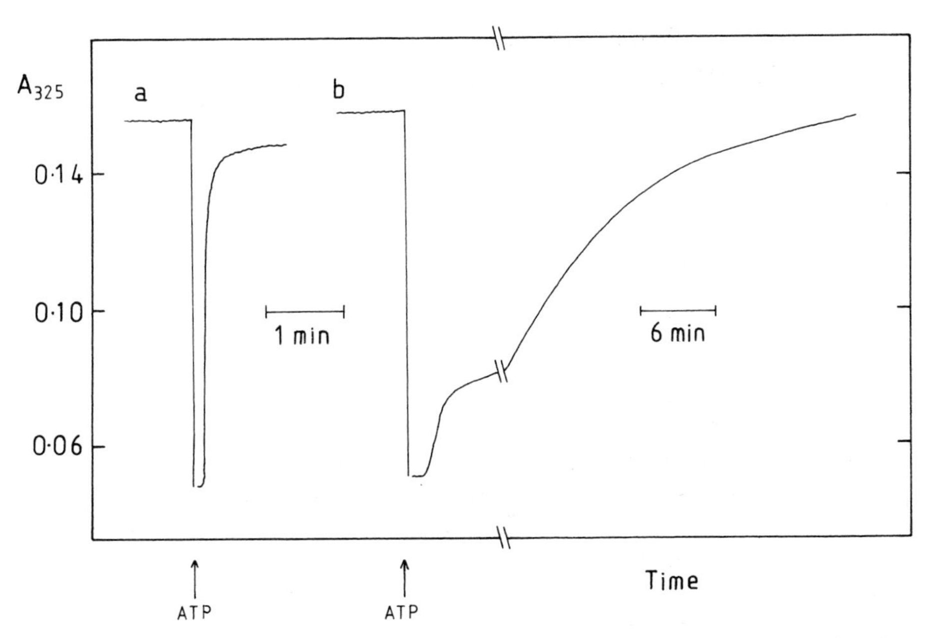

Figure 12. Slow turnover of actomyosin ATPase monitored by turbidity changes. The dissociation of actomyosin, in the presence of ATP, is monitored as a decrease in turbidity (apparent A_{325}) in a spectrophotometer. Trace (**a**) shows the transient dissociation when 4 μM acto-heavy meromyosin (from scallop) turns over 25 μM ATP in the presence of Ca^{2+}. When all ATP is hydrolysed, the myosin recombines with the actin. k_{cat} is estimated as about 1 sec^{-1}. Trace (**b**) shows the same experiment performed in the absence of Ca^{2+}. The recovery phase is now biphasic. This is interpreted as indicating that about 20% of the myosin is unregulated and hydrolyses ATP rapidly (k_{cat} ~1 sec^{-1}) while the remaining 80% hydrolyses ATP very slowly (k_{cat} ~1.8×10^{-3} sec^{-1}). The regulatory mechanism of undamaged actomyosin is therefore displayed *in vitro*, with the high efficiency expected from physiological studies on relaxed muscle. Measurement of ATPase activity alone, however, does not reveal this mechanism, since the 20% of damaged actomyosin hydrolyses most of the ATP during the steady state (effective rate = 0.2 sec^{-1}), while the regulated fraction undergoes a single turnover. [Data from Wells and Bagshaw (19)].

contribution from unmodified enzyme or the true activity of the modified species. Single turnover analysis provides a direct approach to this problem. An analogous situation arises in the study of highly regulated systems such as the ATPase activity of actomyosin from muscle.

6.1 Measurement of Actin-myosin Interaction During ATP Turnover (19)

ATP induces the rapid dissociation of actomyosin, with a concomitant decrease in turbidity. ATP hydrolysis then ensues and when all the ATP is utilised the actin and myosin recombine. ATP turnover is regulated by Ca^{2+}, being very slow in its absence. Changes in turbidity can therefore be used to monitor ATPase activity and reveal the true rate of ATP turnover of the regulated preparation in the absence of Ca^{2+}.

6.1.1 *Method*

(i) Equilibrate 3 ml of 4 μM acto-heavy meromyosin (from scallop adductor muscle) in 20 mM NaCl, 1 mM $MgCl_2$, 10 mM TES at pH 7.5 at 20°C in a cuvette.
(ii) Add 25 μM ATP manually.
(iii) Monitor the turbidity by the apparent A_{325}.

The results are shown in *Figure 12*. In the presence of Ca^{2+}, the dissociation and reassociation phases are fairly rapid (*Figure 12a*). In the absence of Ca^{2+}, however, the reassociation phase is markedly biphasic (*Figure 12b*). The first phase represents the completion of ATP turnover by the unregulated fraction of heavy meromyosin in the preparation, while the slow second phase is due to a regulated fraction which has a highly suppressed ATP turnover rate. The latter, although 80% of the preparation, does not contribute to the steady-state ATP turnover to any significant degree and thus its properties are not revealed by conventional kinetic analysis.

7. SUMMARY

Protein-ligand and protein-protein interactions can be studied by spectroscopic methods, since they are commonly accompanied by a change in protein or ligand fluorescence, by immobilisation of the ligand (giving rise to fluorescence anisotropy) or, more occasionally, by changes in absorbance or light scattering.

Since the molar absorbance of most chromophores lies between 10^2 and 10^4 M^{-1} cm^{-1}, absorption difference spectroscopy is suitable only for procedures where the concentration is varied over the range 10 μM – 1 mM, without the use of specialised apparatus. Fluorescence methods allow lower concentration ranges to be explored but are more prone to artifacts and so require careful control of the conditions (Section 2.2).

Spectrophotometry can be used to determine equilibrium constants for protein-ligand binding. However, in this case, it is vital to establish a proportionality between the signal (optical change) and the amount of bound ligand. This may require calibration using an absolute measurement. Equilibrium methods for the determination of binding constants tend to be limited to $K_d > 1$ μM (Section 4.3) – if binding is tighter than this, the protein and ligand concentrations will have to be decreased below that compatible for reasonable signal to noise. In this case, equilibrium spectrophotometric methods can give accurate values of the stoichiometry of binding (n), but measurement of K_d requires the independent determination of k_{on} and k_{off} for the ligand (Chapter 6).

Applications of spectrophotometry to measurement of rates of processes is also exemplified (Section 6). It is noteworthy that 'transient state kinetics' is not always a term synonymous with 'rapid reaction kinetics' in biological systems. Many reactions, particularly when switched off by a repressor, occur on the time scale of minutes or longer, and make ideal candidates for analysis using conventional spectrophotometers.

7. REFERENCES

1. Lee,A.G. (1978) in *Receptors and Recognition*, Cuatrecasas,P. and Greaves,M.F. (eds.), Chapman and Hall, London, p. 81.
2. Morita,F. (1967) *J. Biol. Chem.*, **242**, 4501.
3. Bagshaw,C.R. (1975) *FEBS Lett.*, **58**, 197.
4. Eccleston,J.F. and Trentham,D.R. (1977) *Biochem. J.*, **163**, 15.
5. White,H.D. and Taylor,E.W. (1976) *Biochemistry (Wash.)*, **15**, 5818.
6. Gaskin,F. (1982) in *Methods in Enzymology*, Vol. **85**, Frederiksen,D.W. and Cunningham,L.W. (eds.), Academic Press, New York, p. 433.
7. Herskowitz,T.T. (1967) in *Methods in Enzymology*, Vol. **11**, Hirs,C.H.W. (ed.), Academic Press, New York, p. 749.
8. Chipman,D.M., Grisaro,V. and Sharon,N. (1967) *J. Biol. Chem.*, **242**, 4388.
9. Eisinger,J. and Lamola,A.A. (1971) in *Methods in Enzymology*, Vol. **21**, Grossman,L. and Moldave,K. (eds.), Academic Press, New York, p. 81
10. Tiedge,H., Lücken,U., Weber,J. and Schäfer,G. (1982) *Eur. J. Biochem.*, **127**, 291.
11. Dandliker,W.B. and Feigen,G.A. (1961) *Biochem. Biophys. Res. Commun.*, **5**, 299.
12. Price,N. and Radda,G.K. (1974) *Biochem. Biophys. Acta*, **371**, 102.
13. Yang,Y.R. and Schachman,H.K. (1980) *Proc. Natl. Acad. Sci. USA*, **77**, 5187.
14. Steiner,R.F. (1984) *Arch. Biochem. Biophys.*, **228**, 105.
15. Bennett,A.J., Patel,N., Wells,C. and Bagshaw,C.R. (1984) *J. Musc. Res. Cell Motil.*, **5**, 165.
16. Holbrook,J.J. (1972) *Biochem. J.*, **128**, 921.
17. Birdsall,B., King,R.W., Wheeler,M.R., Lewis,C.A., Goode,S.R., Dunlap,R.B. and Roberts,G.C.K. (1983) *Anal. Biochem.*, **132**, 353.
18. Ehrenberg,M., Cronvall,E. and Rigler,R. (1979) *FEBS Lett.*, **18**, 199.
19. Wells,C. and Bagshaw,C.R. (1984) *FEBS Lett.*, **168**, 260.
20. Webb,J.L. (1963) in *Enzyme and Metabolic Inhibitors*, Vol. **I**, Academic Press, New York, p.71.

CHAPTER 5

Spectrophotometry and Fluorimetry of Cellular Compartments

C.LINDSAY BASHFORD

1. INTRODUCTION

Spectrophotometry and fluorimetry can be used to monitor intracellular processes, provided that chromophores are present which 'report' on the events in which they participate. It is often fairly straightforward to monitor the optical signals; however, the interpretation of the observations may be extremely problematical. The best approach to simplify interpretation is the correct design of the experiment using the optimal chromophore. *Table 1* lists some systems that can be assessed by fluorimetry or spectrophotometry and indicates which are discussed in more detail below.

It is useful, in the first instance, to distinguish between experiments involving endogenous, or 'intrinsic', chromophores and those using exogenous, or 'extrinsic', pigments provided by the observer. In the former category, studies are usually confined to pigments which absorb light in the visible region of the spectrum, because they are restricted both in number and in the range of biological processes in which they take part. Virtually all biological molecules exhibit absorbance in the ultraviolet but even the most sophisticated spectrometer cannot resolve individual elements of the very complex mixture. In contrast, pigments which absorb in the visible region (porphyrins, flavins, carotenoids) can be distinguished even within tissues and organelles by their characteristic spectra (see Chapter 2) of reasonably narrow, intense absorbance bands. Thus the role of cytochromes in mitochondrial electron transport was appreciated long ago by Keilin (1) as a result of his studies of insect flight muscle using a simple spectroscope. In addition, the co-factors most commonly found in biological oxidation-reduction reactions, namely the pyridine nucleotides (NAD, NADP) and the flavins, fluoresce in their reduced and oxidised forms respectively, and report directly oxidation-reduction reactions in tissues. Photosynthetic systems contain a variety of useful chromophores; chlorophyll fluorescence has been used to monitor light-harvesting ability by photosystems I and II, and carotenoid absorption, as a 'molecular voltmeter', to measure membrane potential, since the position of the absorption peaks alters as a function of electrical potential (ψ) across the membrane (2). Since ψ is a useful parameter in investigating membrane function in other than chloroplast membranes, an extensive search has been made for exogenous pigments (probes), with similar properties to the carotenoids, to use in studying other membrane systems (3).

In addition to exogenous pigments that register membrane potential, dyes have been employed to monitor a wide variety of cellular properties: e.g. cytoplasmic and organelle

Table 1. Systems which can be Assessed by Fluorimetry or Spectrophotometry.

Chromophore	*Method of Measurement*	*Information*
Intrinsic		
Chlorophyll	Fluorescence	Mechanism of
Carotenoid[a]	Absorbance	photosynthesis
NADH	Fluorescence	Electron transport
Flavoprotein	Fluorescence	
Cytochromes[a]	Absorbance	
Haemoglobin[a]	Absorbance	Tissue oxygenation
Myoglobin	Absorbance	
Extrinsic		
Acridines[a]	Fluorescence	pH of cellular
Quene-1[a]	Fluorescence	compartments
Neutral red[a]	Absorbance	
Quin-2	Fluorescence	Cytoplasmic Ca^{2+}
Aequorin	Luminescence	content
Murexide	Absorbance	
Cyanines	Fluorescence	Membrane potential
Oxonols[a]	Fluorescence/absorbance	
Merocyanines	Fluorescence	
Naphthalene sulphonates	Fluorescence	Energy-coupling
		Surface potential
Diphenylhexatriene	Fluorescence	Microviscosity
Anthroyl esters	Polarisation	Membrane fluidity

[a]Experiments involving these chromophores are described in this chapter.

pH, cytoplasmic Ca^{2+} and Mg^{2+} concentration, membrane fluidity or microviscosity, fate of ingested particles. In these cases the molecule in question is usually specifically designed: (i) sensitively to register the parameter of interest and (ii) specifically to reach its ultimate destination. Thus probes of cytoplasmic Ca^{2+} or pH are molecules whose fluorescence or absorbance depends on the cations in their environment. To deliver the indicators, esters are used which are hydrolysed *in situ* by endogenous enzymes to generate the active species precisely in the region of interest.

Because the range of possible probe experiments is so large, it is impractical to describe each in detail. However, it is not always appreciated how simple many of the experiments are to perform using equipment available in most laboratories.

2. APPARATUS

There are three types of instrument usually available for optical experiments: spectrophotometers, which measure absorbance; fluorimeters, which measure fluorescence; and microscopes. The latter do not always provide a photometric recording system although in principle they can be used for fluorimetry or photometry of single cells or small groups of cells. Each apparatus applies a 'geometrical' constraint: fluorimeters and photometers are usually set up for studies of solutions or suspensions in conventional cuvettes (see Chapter 2); microscopes provide two-dimensional images from smears, slices or surfaces. Fluorimeters and photometers can, however, be adapted

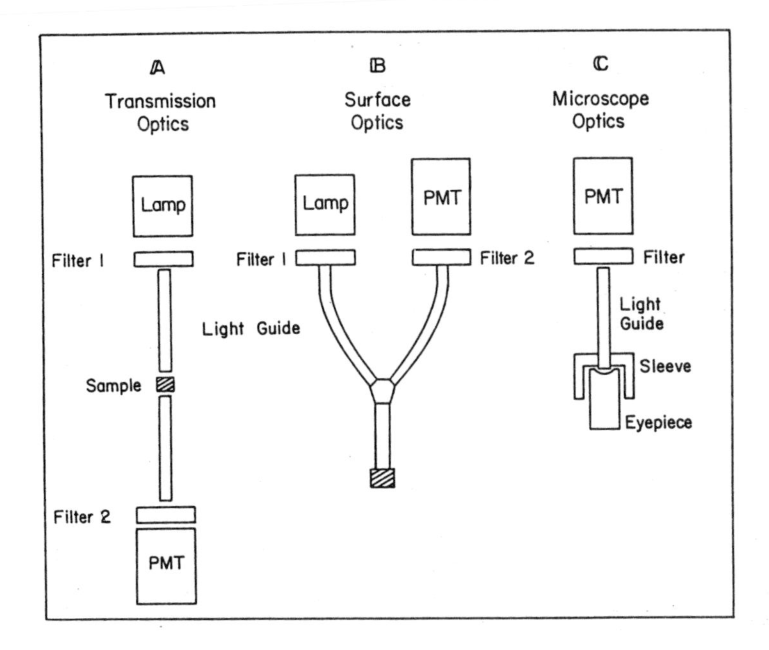

Figure 1. The use of fibre optic light guides to record signals from biological samples. The block diagrams indicate three configurations which we have found useful for recording absorbance or fluorescence signals. **A**. Light is conducted to the sample by a flexible light guide which can be up to 2 m long. Light transmitted through the sample is collected by a second light guide and conducted back to a photomultiplier tube. Wavelength is selected by filters which have rather higher transmission than monochromators. For recording spectra the filters will be replaced by monochromators; for absorbance measurements only one filter is required either in position 1 or position 2; for fluorescence, filter 1 should pass only excitation light and filter 2 only emitted light. *It is very important* to check (despite any manufacturers' claims) that in the absence of fluorophore no signal is recorded! Because of the low intensity of fluorescence emission any excitation light that 'leaks' through the emission filter will normally mask the fluorescence entirely. To avoid stray light other than from the lamp, the apparatus may need to be set up in a dark room, and the sample shrouded to prevent ambient light from reaching the collecting light guide. **B**. In this configuration signals are recorded from the surface of the sample. Similar criteria should be applied as for case **A** above, except that in this case fluorescence and/or reflectance will be recorded. Bifurcated light guides are commercially available with fibres either randomly mixed at the common terminal (the most useful arrangement) or arranged in segments at the common terminal. More complete descriptions of apparatus using light guides for optical coupling can be found in references 13, 24 and 25. **C**. In this case the collecting light guide is held in position over the eyepiece of a microscope. To prevent ambient stray light entering the light guide the fit of the brass sleeve both to the light guide and to the eyepiece should be snug.

to record from surfaces by measuring light emitted and/or reflected from the sample. Front face accessories are available for many fluorimeters, but a convenient method for self-construction uses fibre optic light guides. In this case light from the source is conducted to the surface by a light guide (a bundle of optical fibres) and reflected and fluorescent light is picked up by other fibres and conducted to the detector (usually a photomultiplier; see *Figure 1*).

The reflected light has two components: specularly reflected light, this is the mirror-like reflection, and diffusely reflected light. Only the latter contains information (analogous to absorbance) concerning the chromophores present at the surface. In addition, as in all types of photometry the contributions of stray light (see below) and

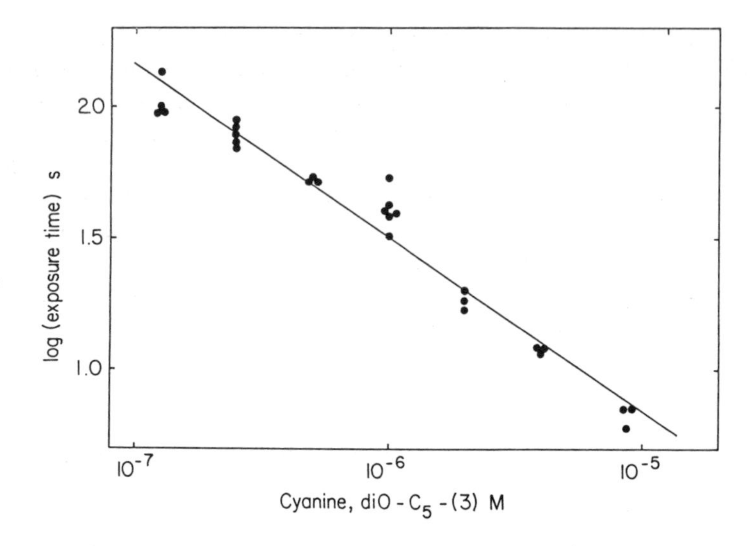

Figure 2. Measurement of cellular fluorescence with a microscope camera. 3T3 cells grown on cover slips were incubated in 2 ml of growth medium containing 2 μl of a stock solution of 3,3′-dipentyloxacarbocyanine (diO-C_5-(3), ref. 26) to give the final concentration indicated. After 2 min the cells were washed twice with phosphate-buffered saline and examined by epifluorescence microscopy using 435 nm excitation and 200-fold magnification. The Leitz Orthomat system was used to measure the time taken correctly to expose a frame of ASA 800 film; the time between the shutter opening and closing was recorded using a stop watch. Fields of view were selected at random using phase contrast optics before recording the fluorescence signal. The exposure time is inversely proportional to dye concentration and thus to fluorescence intensity over two orders of magnitude, brightly fluorescent fields of view yielding the shortest exposures.

the signal of interest have to be assessed by the proper use of filters and/or monochromators.

Light guides can also be used to convert microscopes into photometers or fluorimeters: in this case light from the eyepiece is collected with a light guide and conducted to a suitable detector. We have found that the light guide can be held in place on a microscope simply by a snug-fitting brass ferrule locally machined to provide a tight fit both to the eyepiece and to the light guide. The tight fit excludes room light so that only light from the image is recorded. An alternative procedure for measuring the light output from the image is to take advantage of the automatic light measuring system available on the camera attached to the microscope. In this case the time taken correctly to expose a film of given speed (ASA/DIN) is simply related to the light emitted. Thus, in systems such as the Leitz Orthomat the light meter of the camera will operate even when no film is present and the exposure time can be measured with a stop watch. At very low light levels it is necessary to choose a high ASA setting (high sensitivity) in order for the exposure time to fall into a convenient time range, say 5–120 sec. If the fluorescence is extremely weak the measurement may need to be made in a dark room otherwise stray ambient light will affect the results adversely, registering as apparent fluorescence.

An example of the potential of this approach, in the form of a calibration curve, is shown in *Figure 2*. Cells were stained with different concentrations of cyanine dye and, using filters suitable for use with fluorescein (λ^{ex} = 485 nm, λ^{emit} >520 nm), fluorescence was measured by monitoring the exposure time at ASA 800. There is good correlation between dye concentration and reciprocal exposure time over a very wide range, i.e. exposure time is inversely proportional to the emission intensity. In this technique we used the average exposure time from a number of randomly selected fields (using the phase contrast or bright field mode rather than fluorescence mode) to get a reliable measure of the emission intensity. If photobleaching (see Chapter 1) is severe, as it often can be at high magnifications, it may be necessary to shorten the time that the field is exposed to the fluorescence excitation beam by selecting a higher (more sensitive) ASA setting and only admitting the excitation beam to the sample for the time taken to record the exposure time.

3. EXPERIMENTAL DESIGN

There are three points to consider in the design of optical experiments:

(i) selection of the chromophore;
(ii) characterisation and calibration of the signal;
(iii) checking for artefacts.

3.1. Chromophore Selection

The primary decision is the choice of a suitable chromophore. In the case of intrinsic chromophores this must be a choice of system, namely the one in which the chromophore occurs and in which it exhibits the desired properties. An example is the use of the carotenoid pigments to signal the light-dependent membrane potential associated with photosynthesis. These pigments occur in many photosynthetic membranes but their use as 'molecular voltmeters' is most easily accomplished in preparations, known as chromatophores, from certain photosynthetic bacteria such as *Rhodopseudomonas sphaeroides* or *Rhodopseudomonas capsulata* (see below). On the other hand, studies of oxygen consumption and delivery are best achieved with highly vascularised tissues that contain many mitochondria, such as brain, liver or cardiac muscle; although in the last case the myoglobin contribution cannot be resolved from that of haemoglobin and in all cases the oxygen carriers tend to mask the much less abundant mitochondrial pigments (see below).

For extrinsic chromophores the choice will be constrained by a number of considerations:

(i) can the chromophore be delivered to the desired location?
(ii) will endogenous pigments/factors perturb the response?
(iii) will the chromophore perturb the system under study?

Points (i) and (ii) will be dealt with in the next section. Point (iii) can be tested only by checking the biological properties/activity of the preparation ± added probe.

3.2 Characterisation and Calibration

Once the choice of chromophore and system has been made it is essential to characterise the limits of experimental resolution. In the first instance a study of the spectrum of the chromophore (absorbance or fluorescence) *in situ* may reveal the presence of unexpected contributions to the signal (e.g., different spectra for probe in different cellular environments) and suggest the practical limit of the experiment; in effect this will establish the signal-to-noise ratio for the intended observations (noise being used in its most general sense).

Calibration of particular systems is outlined below (Section 4). It is unfortunately the case that in many experiments the final calibration, be it with pH, transmembrane potential or cation concentration, involves the destruction of the experimental system. Thus to calibrate dye indicators of membrane potential it is usual to modify the potential irreversibly with cations and ionophores. Likewise indicators of cell pH or Ca^{2+} are calibrated at the end of the experiment by permeabilising the cells. Once a suitable calibration procedure is established it is possible to record the desired information from the experimental system.

3.3 **Artefacts**

The problem of artefacts in spectroscopic studies of biological systems is very great. It cannot be emphasised too strongly that very rigorous criteria must be applied before accepting the data at face value. These problems are most acute in fluorescence determinations and a few of the most easily avoided ones are considered here.

3.3.1 *Stray Light*

This is a general term for any light that reaches the detector which is either not of fluorescence origin or, in an absorbance experiment, of a wavelength different from that intended. It may include ambient light and leaks through the monochromators and filters from the source. It is avoided by the correct construction of the fluorimeter or spectrophotometer, the correct choice of blocking filters (to exclude higher order diffraction and specular reflection from the grating in the monochromator) and the appropriate use of dark rooms.

In experiments which involve turbid suspensions, scattered light must be resolved from fluorescence, the latter often being much less in magnitude. Changes in turbidity will affect both fluorescence and absorbance measurements. Substantial changes in turbidity occur when membrane vesicles are exposed to sudden changes in osmolarity or are exposed to permeant salts (4).

3.3.2 *Inadequate Excitation*

Excitation light may be absorbed by non-fluorescent material, as occurs when monitoring the surface fluorescence of tissues or organs. Blood completely absorbs the light used to excite pyridine nucleotide and flavoprotein fluorescence, and hence an increase in the amount of blood in the field of view, due to vasodilation, will decrease fluorescence. This could lead to an erroneous conclusion that the oxidation-reduction state of the tissue mitochondria had changed. To obviate this problem, the ratio of the pyridine nucleotide (PN) and flavoprotein (FP) fluorescence is recorded. As mitochondria become reduced, PN fluorescence increases and FP fluorescence decreases, but changes in exciting light (due to absorption) affect both equally, so their ratio is not sensitive to changes in vasodilation. Hence the FP/PN fluorescence ratio can be used as a sensitive indicator of mitochondrial oxidation-reduction state which is not severely affected by masking problems (5). An alternative procedure for correcting for the interference of blood in

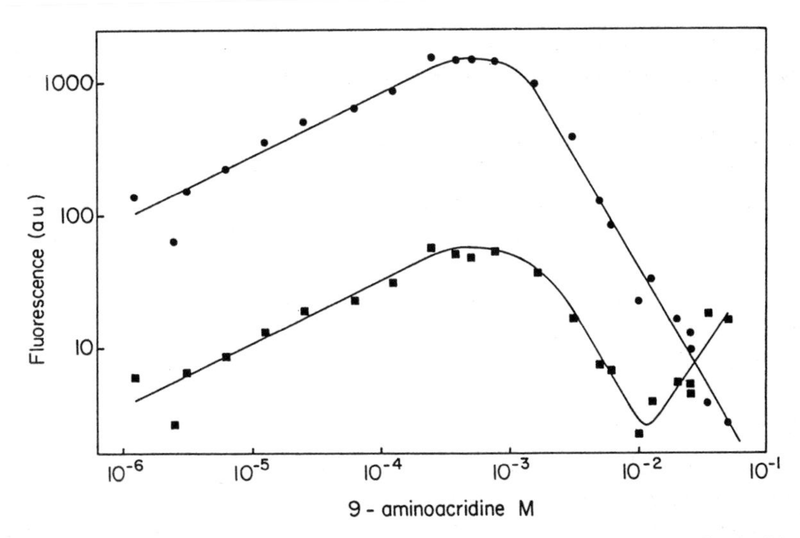

Figure. 3. Concentration dependence of 9-aminoacridine fluorescence. 9-Aminoacridine dissolved in double-distilled water was excited at 400 nm and the fluorescence emission at 456 nm (●) and 550 nm (■) recorded with a Perkin Elmer MPF 44A fluorescence spectrophotometer with excitation and emission slits of 2 nm. To obviate inner filter effects fluorescence was recorded using the 'front face' system with the fluorescent solution contained in a 2 mm pathlength absorbance cuvette whose polished face was set at 60° with respect to the excitation beam. Even with this geometry fluorescence intensity is a non-linear function of fluorophore concentration.

tissue fluorescence is to obtain the 'corrected' fluorescence, namely the difference between the measured fluorescence and the measured reflectance (of the excitation light). In this mode loss of excitation, due to absorbance by the blood pigments, leads to a loss of reflectance which compensates for the loss of fluorescence. The adequacy of this procedure has been demonstrated in perfused systems where the blood can be replaced by a transparent perfusate (6) but its applicability *in vivo* is less certain.

Also, occasionally the excitation light may be absorbed so strongly by the fluorophore itself that fluorescence is reduced along the excitation path, the inner filter effect (see Chapter 1). As a rough rule of thumb the absorbance of the sample (by the fluorophore or any other species), at the excitation wavelength should not exceed 0.05 cm^{-1} fully to avoid inner filter effects and to obtain the expected linear relationship between fluorophore concentration and fluorescence intensity (see Chapter 1 and *Figure 3*). The use of 'front-face' accessories reduces the inner filter effect of highly absorbing samples because the excitation does not have to penetrate very far in order for a fluorescence signal to be recorded. However, filtering will still occur at very high fluorophore concentrations (*Figure 3*).

If a microscope with an epifluorescence attachment is used to excite the material the effective pathlength is of the order of a few microns and little filtering of fluorescence occurs. This can lead to an apparent paradox if results obtained with a microscope (view of surface fluorescence) are compared with those obtained in a conventional fluorimeter (view by transmitted light). Cyanine dyes are avidly accumulated by mitochondria *in vivo* (7) with a resultant quenching of fluorescence when assessed in cell suspensions (8); however, in the fluorescence microscope the mitochondria appear as brightly

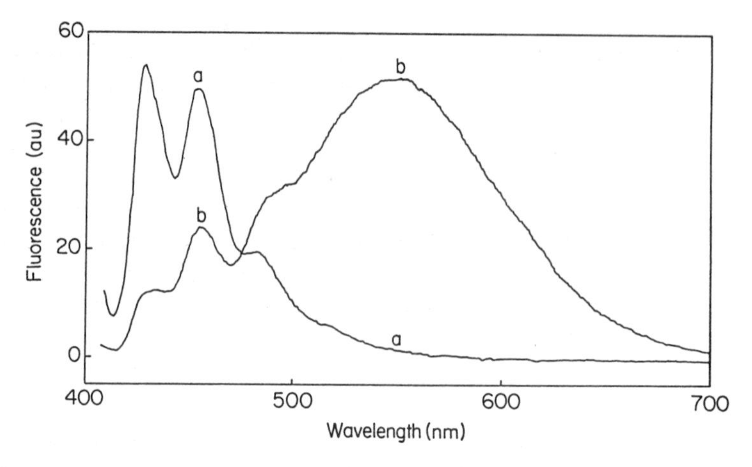

Figure 4. Fluorescence emission spectra of 9-aminoacridine. 9-Aminoacridine dissolved in double-distilled water to give a final concentration of 0.5×10^{-6} M (**a**) or 5×10^{-2} M (**b**) was excited at 400 nm. Emission and excitation slitwidths were 2 nm and the output of the Perkin Elmer MPF 44A fluorescence spectrophotometer was amplified such that the signals at the wavelength of maximum emission were approximately equal. The spectra were scanned at 60 nm/min with an instrument response time of 1.5 sec. The fluorescence maximum of 559 nm is only observed at concentrations of dye exceeding 10 mM (see *Figure 3*).

fluorescent, filamentous structures. Similarly acridine orange and 9-aminoacridine are accumulated by acidic endosomes with a quenching of their blue fluorescence; in the fluorescent microscope the vesicles appear yellow or orange, this being the characteristic emission of very concentrated solutions of these dyes (*Figure 4*, and see *Figure 3*).

3.3.3 *Complex Formation*

The dye chosen may interact with elements of the system with a consequential change in its properties. A pH indicator may not necessarily have the same pK_a when it is bound to cellular constituents as when it is free in solution. In addition the spectrum of the bound dye (in ionised and/or unionised form) may differ from that of the free dye, making any simple evaluation of pH problematical. Alternatively, the dye may be useful as a qualitative indicator of the parameter under investigation but cannot be calibrated because it forms complexes with agents required in the calibration protocol. It is always good practice to use several, independent calibration procedures to check for internal consistency of the dye response. Thus when calibrating optical probes of membrane potential (9,10) both valinomycin/K^+ and uncoupler, carbonyl cyanide *p*-trifluoromethoxyphenylhydrazone (FCCP)/H^+ should give identical values (see below).

4. EXAMPLES

4.1 **Dependence of Mitochondrial Oxidation-reduction State on Tissue Oxygenation**

The properties of mitochondrial suspensions have long been the centre of intense study and the characteristics of the pigments of the respiratory chain have been resolved by dual wavelength spectroscopy (11; see Chapter 2). However, there is always some uncertainty as to whether properties observed in an isolated preparation in a test tube accurately reflect the properties of the unperturbed system *in vivo*. Indeed, on the basis of some spectroscopic observations it was suggested that the terminal enzyme of the mitochondrial respiratory chain, cytochrome oxidase, had a much lower affinity for oxygen in the brain *in vivo* than it did in isolated preparations of mitochondria (12). One way to resolve the controversy is to measure the absorbance spectrum of brain *in vivo* and see when the elements of the respiratory chain become reduced as the oxygen tension of the tissue is lowered. This can be achieved by recording the diffuse reflectance spectrum of a suitably exposed tissue. In this particular set of experiments the reflectance spectra were recorded from freeze-trapped specimens at liquid nitrogen temperatures to avoid temporal distortions of the optical readout and to enhance the spectral resolution (13,14).

4.1.1 *Method*

Anaesthetised gerbils breathed gas mixtures of defined composition (0−95% O_2, 5% CO_2 with the balance as N_2) which were delivered by a gas mixer to a mask which covered the face; either a plastic funnel or a large syringe with the plunger removed make excellent masks for this purpose. After 60 sec their brains were freeze-trapped using a surface freezing procedure (15). A funnel was fitted into an incision that exposed the skull and liquid N_2 was poured into the funnel until respiration ceased (this took about 60 sec). Care must be taken not to allow nitrogen from the freezing funnel to mix with the respiratory gases in the mask. Finally the whole animal was immersed and stored in liquid N_2. The brain was exposed by mechanical milling in an environment of liquid N_2 and reflectance spectra were recorded from the milled surface. The baseline was the white polystyrene container (a conventional white ice box) that contained the frozen head in liquid N_2.

The reflectance spectra were recorded as follows: the head was held in a clamp in liquid N_2; the common terminal of the sampling light guide (see *Figure 1*) was cooled in the liquid N_2 and then gently held on the exposed brain surface. Contact between the light guide and the sample minimized the specular reflection. The spectrum was scanned and the data was stored in the computer which controlled the spectrophotometer; the baseline was recorded over the same wavelength range and finally the reflectance spectrum was plotted (as an absorbance spectrum) as the difference between the sample and the baseline scans.

4.1.2 *Results*

A series of such spectra are shown *Figure 5* (traces 1, 2, 3). The principal features of the spectrum between 500 and 650 nm arise from haemoglobin (maximum at 555 nm), oxyhaemoglobin (maxima at 542 and 577 nm) and from reduced mitochondrial cytochromes. If the animal was breathing 95% O_2 (trace 1) only peaks due to oxyhaemoglobin are found; if the animal was breathing 5% O_2 then the peak due to haemoglobin is found; only when the animal was switched to 0% O_2 are peaks due to ferrocytochromes $c+c_1$ and aa_3 clearly resolved (trace 3). Note that the animal breathing 5% O_2 still has some residual oxyhaemoglobin detectable as shoulders on the haemoglobin spectrum (trace 2). Since there is little oxyhaemoglobin present when the animals breathed either 5% or 0% O_2 the difference between these spectra (*Figure 5*, trace 4) is almost identical to that found for oxidised *minus* reduced mitochondrial

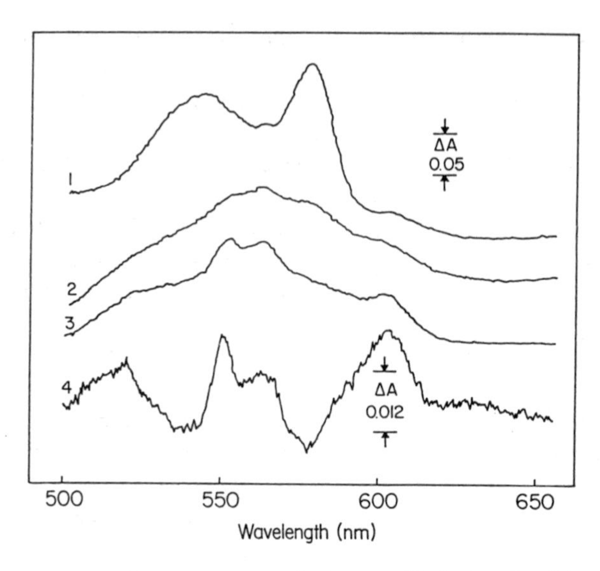

Figure 5. Reflectance spectra of freeze-trapped cerebral cortex. The brains of gerbils were freeze-trapped (15) 1 min after the anaesthetised animal began breathing: (**1**) 95% O_2, 5% CO_2; (**2**) 5% O_2, 90% N_2, 5% CO_2; (**3**) 95% N_2, 5% CO_2. Reflectance spectra **1**–**3** were recorded using a scanning spectrophotometer fitted with a bifurcated light guide (see *Figure 1B*) from the right cortex at 77K. The baseline was recorded from the white, polystyrene container (ice bucket) containing the frozen preparation. The difference spectrum (**4**) represents spectrum **3** minus spectrum **2**. Reproduced with permission from reference 27.

suspensions; the peak at 550 nm arises from ferrocytochromes $c + c_1$ and that at 605 nm arises from ferrocytochromes aa_3; the minor peak at 565 nm may arise from residual haemoglobin or from ferrocytochromes *b*.

4.1.3 *Conclusions*

It is clear from this sequence of spectra that as the animals passed from breathing 95% O_2 almost all the oxyhaemoglobin absorbance disappeared without any concomitant appearance of ferrocytochrome peaks. Only in extreme anoxia did reduction of the mitochondrial pigments occur. This shows that the respiratory chain still has enough oxygen under conditions where tissue haemoglobin is almost completely disoxygenated; or, put another way, the large difference in affinity for oxygen exhibited by mitochondria and haemoglobin *in vitro* is still apparent *in vivo*.

The experiments described here were performed with a locally constructed, microprocessor controlled scanning spectrophotometer which was capable of storing and manipulating spectra. However, commercial instruments with similar capacity for reflectance measurements have recently been introduced, viz. the Perkin Elmer Lambda 5 UV/vis Spectrophotometer equipped with an External Integrating sphere, and a custom built double wavelength spectrometer from Applied Photophysics Ltd., London.

4.2 Measurement of Membrane Potential in Organelles and in Cells

Most biological membranes have the ability to sustain electrostatic potential differences between the aqueous phases which they separate. This potential, usually called the membrane potential, is often a significant component of membrane function; it is, for example, critical for signalling in excitable cells and makes a contribution to the accumulation of nutrients driven by the electrochemical Na^+ gradient across the plasma membrane of many cells. In subcellular organelles such as chloroplasts and mitochondria the membrane potential is an important 'high energy' intermediate in the synthesis of ATP.

In most of these cases it is impractical to record the potential directly with microelectrodes, and of the other methods available for measuring transmembrane potential the use of dyes with specific sensitivity to potential is one of the most popular (16). While very many dyes change either their absorbance or their fluorescence as a response to a change in membrane potential (and thus are potentially useful probes) (3), it is often difficult accurately to calibrate the signal with known potentials. The most widely available procedure involves modification of membrane permeability to a particular cation, usually K^+ or H^+, with a reagent called an ionophore. In the presence of sufficient ionophore, such that all other permeabilities are relatively small, the membrane potential (V) will approach that predicted by the Nernst equation:

$$V = (-RT/F)\ln([M^+]_{in}/[M^+]_{out}) \qquad \text{Equation 1}$$

where M^+ is the relevant ion, R is the gas constant, T is the absolute temperature and F is Faraday's constant. At 37°C this relation simplifies to:

$$V(mV) = -61.5 \log [M^+]_{in}/[M^+]_{out} \qquad \text{Equation 2}$$

which shows that for a 10-fold change in cation gradient a 61.5 mV change in potential is obtained. The cation gradient can be manipulated by altering the medium concentration of M^+, and hence a calibration curve of optical signal and potential can be obtained in the presence of ionophore.

4.2.1 *Light-induced Membrane Potential in Chromatophores*

Chromatophores prepared from photosynthetic bacteria provide an ideal model system for the study of the behaviour of membrane potential-indicating dyes. Electron transport in the photosynthetic reaction centre is initiated by infra-red light, and short intense flashes can be used to turn the photosynthetic apparatus just once ('single turnover' flashes). On electron transport, protons are pumped into the chromatophore with the generation of a substantial membrane potential (up to 300 mV inside positive). Furthermore the chromatophore membrane contains pigments, the carotenoids, whose absorbance position varies directly with the amplitude of the potential. The potential-dependent shift of carotenoid absorption can be calibrated with K^+ in the presence of valinomycin (ref. 2; see below for a full description of this procedure). *Figures 6–8* illustrate the complementary use, in the *R.sphaeroides* system, of an endogenous chromophore (carotenoids) and an exogenous 'probe' (oxonol dye, *Figure 9*). *Figure 6* shows the behaviour of the added dye. Its spectrum in the dark (trace a) was obtained by subtracting the spectrum of chromatophores (in the dark) from that of chromatophores plus dye; similarly the absorbance spectrum of the dye in illuminated chromatophores (by a projector bulb through an infra-red filter) was recorded (trace b). Clearly illumination of the chromatophores causes a substantial red shift of oxonol absorbance. This light-dependent shift is abolished in the presence of uncoupling agents or in the

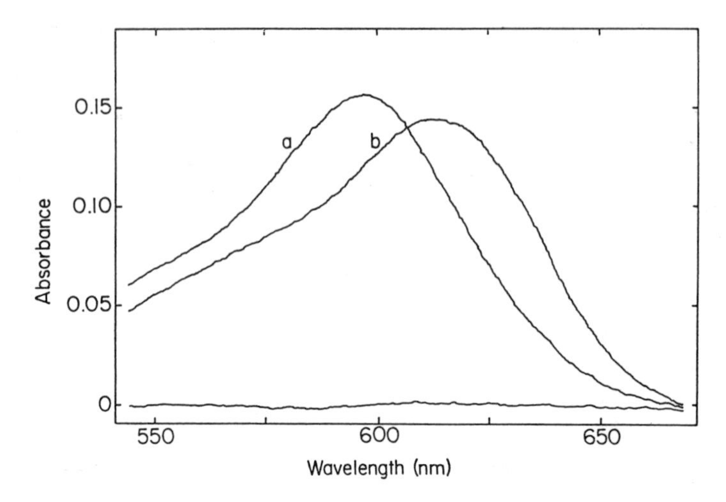

Figure 6. The absorbance of oxonol-VI incubated with chromatophores prepared from *R.sphaeroides*. The absorbance of chromatophores (21×10^{-6} M bacteriochlorophyll) suspended in 100 mM KCl, 20 mM morpholinopropane sulphonic acid (MOPS), 1 mM $MgCl_2$, 0.5 mM ascorbate, pH adjusted to 6.9 with KOH at 23°C in the absence (dark) or presence (light) of light from a 15 W tungsten filament filtered through a Kodak Wratten Gelatin filter, number 88A, were stored as baselines in the digital memory of a scanning spectrophotometer. Oxonol-VI (*Figure 9*) was added to give a final concentration of 1.5×10^{-6} M and **spectrum a** was recorded in the dark and **spectrum b** in the light. Reproduced with permission from reference 17.

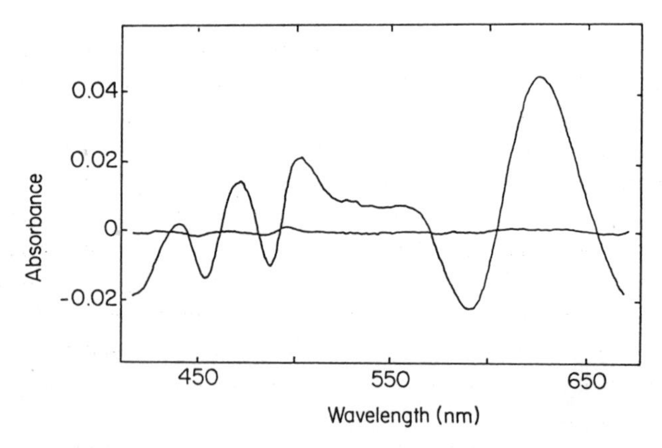

Figure 7. The light-dark difference spectrum of *R. sphaeroides* chromatophores in the presence of oxonol-VI. Chromatophores and oxonol-VI were incubated as described in the legend to *Figure 6*. The spectrum of the suspension in the absence of illumination from the tungsten filament was used as the baseline. Reproduced with permission from reference 17.

presence of valinomycin and excess K^+, agents which prevent the formation of a membrane potential by proton pumping.

Figure 7 illustrates a similar experiment in which the baseline was the spectrum of the chromatophores plus dye in the dark. The light *minus* dark difference spectrum has a trough around 590 nm and a peak around 630 nm which arise from the red shift of the oxonol absorbance (this region is featureless in the absence of dye at this sensitivity), and a series of peaks and troughs between 400 and 520 nm which arise

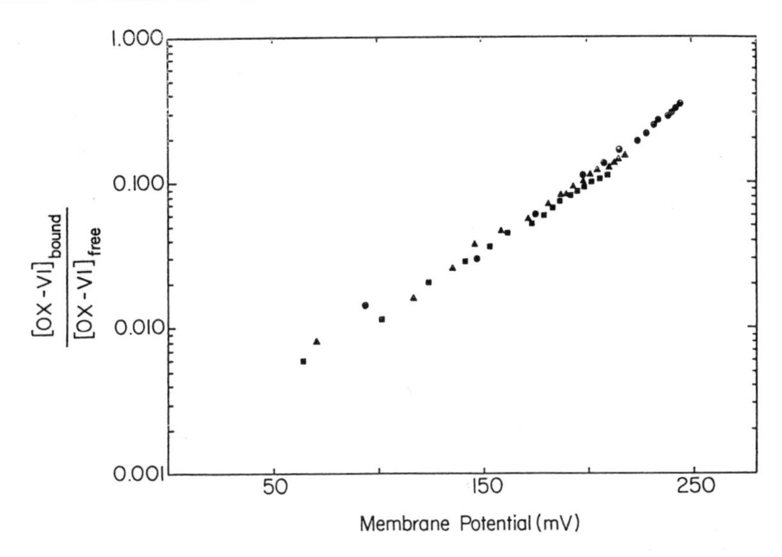

Figure 8. The membrane potential-dependent binding of oxonol-VI to chromatophores of *R. sphaeroides*. Chromatophores (19.4×10^{-6} M bacteriochorophyll) were suspended in a medium similar to that described in the legend to *Figure 6* containing 3×10^{-6} M (●), 5.9×10^{-6} M (▲) or 8.7×10^{-6} M (■) oxonol-VI. The membrane potential during illumination was estimated from the absorbance change at 490−475 nm [the carotenoid bandshift (2,17) calibrated using valinomycin *plus* K^+, see *Figures 10* and *11*] and the amount of oxonol-VI bound under the same conditions was calculated from the absorbance change at 587 nm (17). Reproduced with permission from reference 17.

from the red shift of carotenoid absorbance. Thus it is possible to record signals from both the intrinsic and the extrinsic chromophores together. Behaviour of the former signals can thus be used to understand the properties of the latter.

Calibration of the oxonol signal was achieved by an indirect procedure: the response of the carotenoids to pulses of K^+ in the presence of valinomycin was first monitored and then the behaviour of the oxonol was compared with that of the carotenoids. The reason for this approach was that in these small vesicles it proved impossible to maintain a steady potential with valinomycin *plus* K^+ in the presence of oxonol. Having calibrated the system, measurements of potential were then possible under illumination with light of different intensities or by altering the frequency of single turnover flashes. The results shown in *Figure 8* indicate that the dye has a logarithmic response to potential in this system (as measured by carotenoid response).

A detailed study of this system (17) revealed that the mechanism by which the oxonol responded to potential was by migration into the chromatophore milieu (the distribution of the dye across the membrane obeys Equation 1) and subsequent binding of the dye to hydrophobic sites. Many of the most useful optical indicators of membrane potential work by this accumulation *plus* binding mechanism.

4.2.2 *Plasma Membrane Potential of Animal Cells*

The membrane potential of whole cells is more difficult to assess with optical indicators than that of isolated organelles because the indicator can distribute among many subcellular compartments. It is necessary to isolate the response of the plasma membrane from that of any other cell membrane. For example, the intense staining of mitochondria

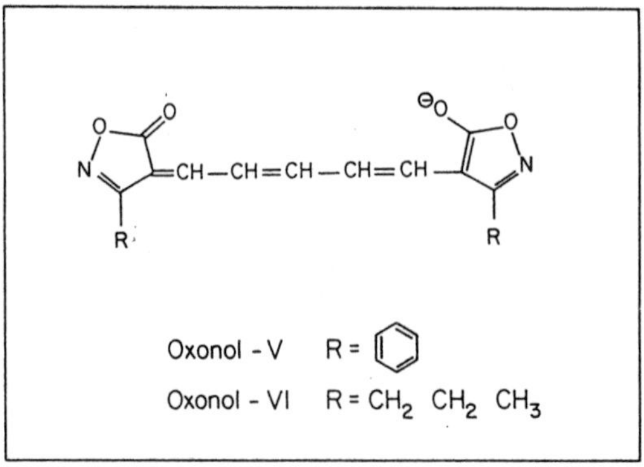

Figure 9. The structure of membrane potential-indicating oxonol dyes. Oxonol-V is bis[3-phenyl-5-oxo-isoxazol-4-yl] pentamethineoxonol (28) and oxonol-VI is bis [3-propyl-5-oxoisoxazol-4-yl] pentamethine oxonol. Details of their synthesis are described in reference 28. These, and many other useful optical indicators are available from Molecular Probes Inc., 24750 Lawrence Road, Junction City, OR 97448, USA.

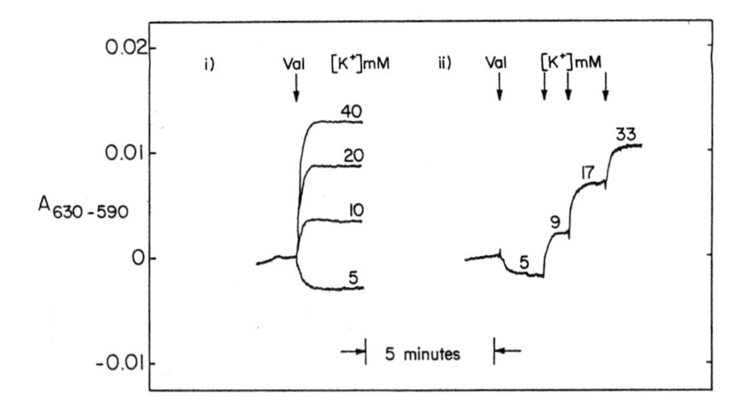

Figure 10. Measurement of Lettre cell membrane potential using oxonol-V. (**i**) 5×10^6 Lettre cells/ml were suspended in 5 mM Hepes, 1 mM $MgCl_2$, 2.2×10^{-6} M oxonol-V, pH adjusted to 7.4 with NaOH at 33°C and 155 mM NaCl *plus* KCl with the final K^+ concentration indicated. Valinomycin was added as indicated to give a final concentration of 1×10^{-6} g/ml. Traces from four separate experiments are superimposed such that the signal before the addition of valinomycin is identical. The extracellular concentration of K^+ at which valinomycin would give no change in absorbance was estimated to be 7.1 mM by interpolation, and the intracellular K^+ concentration was 73.3 mM giving a membrane potential (see Equation 2) under these conditions of −62mV. (**ii**) 5×10^{-6} Lettre cells/ml were suspended in 150 mM NaCl, 5 mM KCl, 5 mM Hepes, 1 mM $MgCl_2$, 2.2×10^{-6} M oxonol-V, pH adjusted to 7.4 with NaOH at 33°C. Valinomycin (1×10^{-6} g/ml) and KCl to give the concentration indicated were added as shown by the arrows. The extracellular K^+ concentration at which valinomycin would give no change in absorbance was 6.9 mM, the cellular K^+ concentration was 63.3 mM and thus the membrane potential was −60 mV. Reproduced with permission from reference 10.

by cyanine dyes (7), which are membrane-permeant cations, requires that, to measure plasma membrane potential using a cyanine dye, or other lipophilic cation, conditions must be established where the mitochondrial potential is disabled. If an anionic dye is chosen this problem is overcome, but in this case there is very little entry of dye into the cells unless the dye is itself reasonably lipophilic. Thus for studies of whole

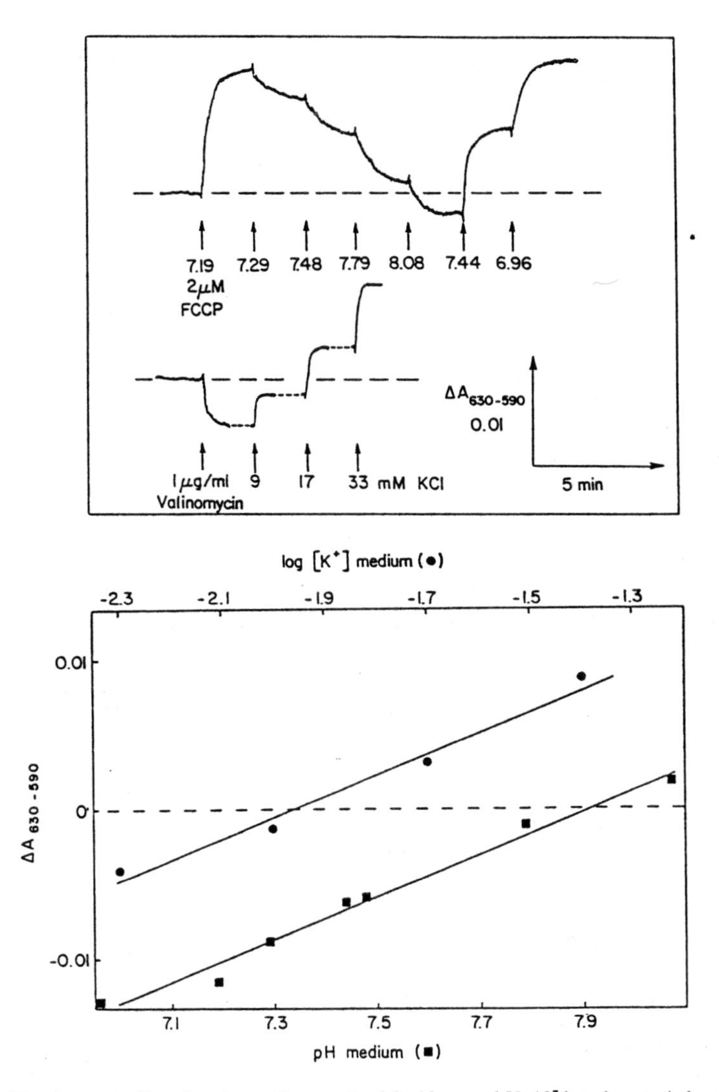

Figure 11. Measurement of lymphocyte membrane potential with oxonol-V. 10^7 lymphocytes/ml were suspended in 150 mM NaCl, 5 mM KCl, 5 mM Hepes, 1 mM $MgCl_2$, 2×10^{-6} M oxonol-V, pH adjusted to 7.4 with NaOH at 37°C. FCCP and NaOH or HCl (upper panel, upper trace) or valinomycin and KCl (upper panel, lower trace) were added to give the final concentration or pH indicated. Lower panel: the difference between the absorbance recorded before and after the addition of FCCP at the relevant pH values (■) or of valinomycin at the relevant KCl concentrations (●) are plotted on equivalent axes. The concentration of H^+ or K^+ at which the addition of FCCP or valinomycin, respectively, would have given no change in absorbance is obtained from the point of intersection of the solid lines with the baseline (dashed). In these experiments the K^+ 'null-point' was 11 mM and the pH null point was 7.91: intracellular K^+ was 102 mM and intracellular pH was 7.0 giving membrane potentials of −59 and −56 mV, respectively. Reproduced with permission from reference 10.

cells we employ an oxonol dye similar to that used in the chromatophore studies but with phenyl rather than propyl substituents (*Figure 9*). The procedure for measuring plasma membrane potential with this dye, oxonol-V is given below (*Figures 10* and *11*):

(i) Suspend the cells in a physiological saline in the presence of dye at 37°C, and equilibrate for about 10 min, after which the absorbance value should be steady. (Do not use serum, or other albumin-containing medium for suspension, because the albumin will sequester the dye, and none will interact with the cells.)
(ii) Alter the permeability of the cell membrane by addition of an ionophore, and note the change in dye absorbance.
(iii) Determine the effect of ionophore on dye absorbance at different extracellular concentrations of the cation for which the ionophore is specific.

Whole cells have, in general, a sufficiently large reservoir of internal K^+ to maintain a steady potential in the presence of oxonol and valinomycin, and thus the change in absorbance levels off (*Figures 10* and *11*).

In principle, no change is observed if the K^+ equilibrium potential (diffusion potential) equals the pre-existing membrane potential – the K^+ 'null point'. Under these conditions, the membrane potential is given by Equation 1, with the external K^+ concentration = $[K^+]_{out}$ and the internal K^+ concentration = $[K^+]_{in}$. Determination of $[K]_{in}$ is described below, and since $[K^+]_{out}$ is known, the membrane potential can be calculated.

Normally, of course, some change in dye absorbance will be observed on addition of valinomycin, since $[K^+]_{out}$ is not at the correct value to match the pre-existing membrane potential (*Figure 10*). Two approaches are then possible. First, the cells may be incubated in media of various K^+ concentrations before valinomycin addition, and the null point found by interpolation between the closest values of $[K^+]_{out}$ (*Figure 10*). This procedure has the disadvantage that it uses one sample of cells for each value of $[K^+]_{out}$. In addition it works only with cells of rather low K^+ permeability; cells fairly permeable to K^+ depolarise in high K^+ media, and subsequent addition of valinomycin has little effect on potential.

More convenient is the procedure shown in *Figure 11*. In this case, a fairly low value of $[K^+]_{out}$ is initially used and, after valinomcyin addition, the concentration of $[K^+]_{out}$ is increased by addition of small aliquots of concentrated KCl (4 M is about the highest manageable concentration). As $[K^+]_{out}$ is increased, the null point is crossed and, again, its value can be determined by interpolation. This method uses a single sample of cells for the complete titration, and has the advantage of speed and economy of use of the biological preparation. In both cases, it is usually sufficient to interpolate the data points by eye – sophisticated least squares analysis is unnecessary, especially since the theoretical form of the 'absorbance' versus 'potential' plot is unknown.

In either procedure a knowledge of cytoplasmic K^+ concentration is required to calculate the potential accurately. This can be achieved by pelleting the cells through oil and analysing the water (wet–dry weight) and K^+ content of the pellet. A good oil for such a determination is a mixture of di-n-butylphthalate (two parts) and dinonylphthalate (one part) with a density of 1.02; cells pellet through such an oil within 10 sec in a Beckman microfuge B. Cell cations can be determined (*after* the water content!) by atomic absorption spectrometry. The spillover of medium into the cell pellet is usually negligible, less than 0.05%, but this can be verified by including an extracellular space marker, such as [^{14}C]inulin, in the incubation medium.

There is a danger that the lipophilic valinomycin/K^+ complex may precipitate

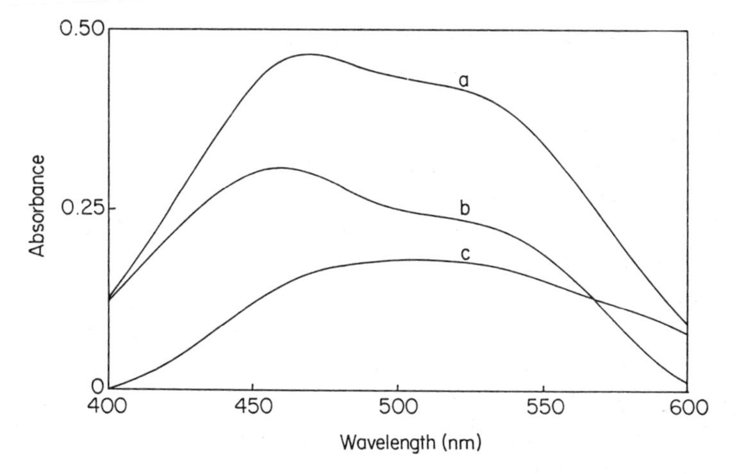

Figure 12. Measurement of Lettre cell pH with neutral red. The absorbance spectra of four samples were recorded at 23°C and stored in the memory of a scanning spectrophotometer: (**1**) medium alone comprising 150 mM NaCl, 5 mM KCl, 5 mM Hepes, 1 mM $MgSO_4$, pH adjusted to 7.3 with NaOH; (**2**) medium plus 3×10^{-6} Lettre cells/ml; (**3**) medium plus 3×10^{-6} Lettre cells/ml plus 2.5×10^{-5} M neutral red; (**4**) supernatant remaining from medium **3** after pelleting the cells at 16 000 $g \times 1$ min in a Beckman microfuge B. The following difference spectra are presented: **trace a**: scan 3 − scan 2, the spectrum of extracellular plus intracellular neutral red; **trace b**: scan 4 − scan 1, the spectrum of extracellular neutral red; **trace c**: (scan 3 − scan 2) − (scan 4 − scan 1), the spectrum of intracellular neutral red, that is the difference between **spectrum a** and **spectrum b**. The pH of the intracellular and extracellular compartments was calculated from the ratio of the absorbance at 530 nm to that at 477 nm. For **trace b**, extracellular space, A_{530}/A_{477} = 0.768, pH = 7.32; for **trace c**, intracellular space, A_{530}/A_{477} = 1.032, pH = 7.01.

lipophilic anions such as the oxonol dyes and it is advisable to check on the calibration described above with another ionophore. The uncoupling agent FCCP is useful in this respect as it selectively increases the H^+ permeability of membranes. The experimental protocol for measuring the potential using FCCP is identical to that using valinomycin except that medium pH rather than K^+ is varied by addition of acid or base (*Figure 11*). Calculation of the membrane potential using the pH null point requires a knowledge of cell pH. This is usually close to 7.1 but may have to be determined separately either by using indicators (see below) or by some other experiment.

4.3 Measurement of the pH of Cellular Compartments

4.3.1 *pH of Endosomes*

Most eukaryotic cells internalise components of their plasma membrane and associated exogenous material by endocytosis. The subsequent fate of the endosomes may be complex but it usually involves a step in which the vesicle milieu is acidified (18). The pH of such an endosome can be monitored directly if a pH-sensitive dye can be attached either to the relevant piece of membrane or to a suitable extracellular vehicle which is endocytosed. The dye most commonly used is fluorescein because it is easily visualised with a fluorescence microscope and because its isothiocyanate derivatives can be covalently conjugated to vehicles such as proteins (for receptor-mediated endocytosis) or dextran. Furthermore the fluorescence of fluorescein and its conjugates is sensitive to pH in the range 5.5−7.5 (19). The endosomes can be loaded with the fluorescein

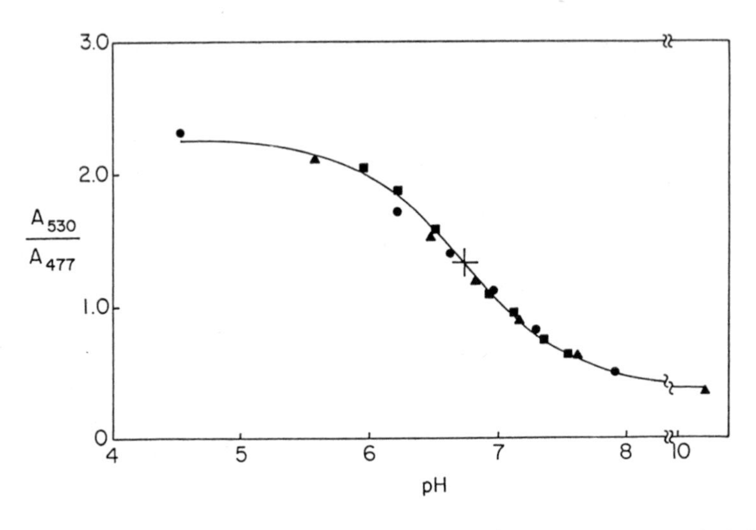

Figure 13. pH dependence of neutral red absorbance. 1.7×10^{-5} M (●,▲) or 9×10^{-6} M (■) neutral red dissolved in 0.05 M KH_2PO_4 at 23°C with pH adjusted as indicated with NaOH. Absorbance was recorded at 530 nM, the absorbance maximum at low pH values, and 477 nm the isosbestic point (see Chapter 2). The ratio A_{530}/A_{477} is presented as a function of pH and the solid line is the best fit Henderson-Hasselbalch curve which has the form:

$$\mathrm{pH} = 6.737 + \log \frac{2.267 - A_{530}/A_{477}}{A_{530}/A_{477} - 0.385}$$

Where the acid 'end point' is 2.267, the base 'end point' is 0.385 and the pK_a is 6.737.

derivative by incubating the cells in a suitable physiological medium containing the dye or labelled molecule. The extracellular dye is subsequently removed by washing.

For cells attached to a solid substrate such as glass or plastic, fluorescence spectra can be recorded in a conventional 1 cm pathlength cuvette provided that the substratum is made such that it can be fitted diagonally into the cuvette (about 4 cm ×1 cm × 0.1 cm). It is always much easier to shape the substrate before attaching the cells to it. With the cells in the cuvette in a suitable medium fluorescence spectra can be recorded directly. Calibrate the signal as follows:

(i) Add the ionophore monensin to a final concentration of 1 μg/ml: this will equilibrate the pH throughout the cell and the medium.

(ii) Titrate the medium until the spectrum of the dye in the absence of ionophore is restored: the medium pH value now equals that originally present in the endosome.

(iii) Check that the label remains within the cells at the end of the experiment, i.e. that the ionophore does not cause leakage of the label into the medium. To do this, remove the cells and substratum and record the spectrum of the remaining buffer.

(iv) In addition, inspect the removed cells for lysis with a fluorescence microscope.

Problems with this technique are unwanted autofluorescence of cells which increases on cell death, and the possibility that the fluorescence of the chromophore concentrated in a small compartment may not be identical to that found by the same amount of material in dilute solution.

A more qualitative estimate of endosome pH can be obtained by monitoring the distribution of a permeant, fluorescent amine, such as 9-aminoacridine or acridine orange, whose fluorescence varies with concentration. Such amines accumulate in acid compartments because the protonated form of the dye is much less membrane permeant than the free base. Indeed if the latter equilibrates across membranes then the pH within the vesicle is given by the relationship (20):

$$pH_{in} = pH_{out} - \log ([AH^+]_{in}/[AH^+]_{out}) \qquad \text{Equation 3}$$

Normally $pH_{out} < pK - 1$ so that $[AH^+] \simeq [AH^+] + [A]$. When present at concentrations exceeding 10 mM, 9-aminoacridine exhibits a yellow rather than a blue fluorescence (*Figure 4*) and acridine orange a red fluorescence. Cells incubated with 0.01 mM 9-acridine orange have many yellow vesicles when examined by fluorescence microscopy using the fluorescein filters; this indicates that such vesicles have accumulated the dye by at least two orders of magnitude and that the internal pH must be in the range 5–6.

4.3.2 *pH of Cytoplasm or Isolated Membrane Vesicles*

As with endosomes, it is possible to monitor cytoplasmic pH by trapping an appropriate dye in the cell cytoplasm. For this purpose indicators of a similar type to that used for cytoplasmic Ca^{2+} measurements are convenient (21). We have used an indicator known as quene-1 (22).

(i) Load the dye into cells by adding the acetoxymethyl ester of the dye outside the cells. This permeates the plasma membrane and is subsequently hydrolysed by cellular esterases to the impermeant acid form.

(ii) Follow the progress of hydrolysis by recording the progressive change in spectrum from that of the ester to that of the acid. It is important to include bicarbonate in the incubation medium as the neutral esterases in the cytoplasm are much less active at acid pH values and hydrolysis of the ester form of the indicator generates intracellular acid.

(iii) After loading, wash the cells several times until the medium fluorescence is reduced to a minimal value. The remaining signal arises from the cytoplasmically located dye.

(iv) Calibrate the signal by permeabilising the cells at the end of the experiment with 0.5% Triton X-100 in the presence of 0.5 mM EGTA and 0.5 mM EDTA to chelate metal ions, which are potential quenchers of fluorescence, and titrate the suspension until the fluorescence observed in the absence of Triton is regained (cf. above).

A word of caution about the cleaning of cuvettes is appropriate here; chromic acid may introduce fluorescence quenchers; these can usually be removed by soaking the cuvettes in solutions of chelating agents such as EDTA.

Unfortunately, some cells seem not to retain the acid form of these pH and Ca^{2+} dyes very long even though the free acid is supposed to be membrane impermeant. Again, it is very important to check that the signal observed does indeed originate from within cells; this is best done by pelleting a sample of the suspension and verifying that the supernatant lacks dye. If leakage is particularly troublesome, reasonable data can be obtained by loading the cells for only a short time (much less than the time taken for complete hydrolysis of the ester), rapidly washing the cells and immediately making the pH or Ca^{2+} determination. Unfortunately this leads to an inevitable wastage of rather an expensive starting material.

An alternative to using an entrapped pH indicator is to use a permeant indicator and then to correct for the contribution from extracellular dye. Neutral red is ideally suited for this purpose, as it readily enters cells, and provided that external buffering is strong any perturbations of the signal can be ascribed to pH changes in the closed compartment (23).

A typical experiment, in which the absolute value of cellular pH was monitored with neutral red, is shown in *Figure 12*. Lettre cells (an ascites tumour cell line) were incubated with physiological Hepes-buffered saline in the presence of 0.025 mM neutral red at room temperature and the spectrum recorded (trace a). With 5×10^6 cells/ml approximately 50% of the dye is cell associated. The cells were then pelleted and the spectrum of the supernatant recorded (trace b). Trace c is the difference between the two former spectra, that is the spectrum of cell-associated dye. Again the technique makes use of the ability of modern spectrometers to store spectra for subsequent analysis as the spectra presented required four scans:

(i) a scan of medium alone, the baseline for trace b;
(ii) a scan of cells plus medium, the baseline for trace a;
(iii) a scan of cells plus medium plus dye;
(iv) a scan of medium plus dye after pelleting the cells.

In the figure trace a is scan (iii) *minus* scan (ii), trace b is scan (iv) *minus* scan (i), and trace c is the difference between the difference spectra a and b.

The pH of each compartment was calculated from a standard curve established using solutions of neutral red of known pH. The parameter chosen was the ratio of the absorbance due to the acid form (A_{530}) to that of the isosbestic wavelength (A_{477}) since this obeys the Henderson-Hasselbalch relationship and is independent of dye concentration (*Figure 13*). The pH of Lettre cells under these conditions was found to be 7.01. In this calculation, we assumed that the pK_a and the spectral properties of cell-associated dye are identical to those of free dye, an assumption that remains to be validated. However, the value obtained by the indicator technique described here agrees well with that obtained by two independent techniques (24).

5. ACKNOWLEDGEMENTS

I would like to thank Mrs B.J.Bashford for her patient preparation of the diagrams and our own experiments were supported by the Royal Society and the Cell Surface Research Fund.

6. REFERENCES

1. Keilin,D. (1925) *Proc. R. Soc. Lond. B*, **98**, 312.
2. Jackson,J.B. and Crofts,A.R. (1969) *FEBS Lett.*, **4**, 185.
3. Cohen,L.B., Salzberg,B.M., Davila,H.V., Ross,W.N., Landowne,D., Waggoner,A.S. and Wang,C.H. (1974) *J. Membrane Biol.*, **19**, 1.
4. Bangham,A.D., De Gier,J. and Greville,G.D. (1967) *Chem. Phys. Lipids*, **1**, 225.
5. Chance,B., Schoener,B., Oshino,O., Itshak,F. and Nakase,Y. (1979) *J. Biol. Chem.*, **254**, 4764.
6. Harbig,K., Chance,B., Kovach,A.G.B. and Reivich,M. (1976) *J. Appl. Physiol.*, **41**, 480.

7. Johnson,L.V., Walsh,M.L., Bockus,B.J. and Chen,L.B. (1981) *J. Cell Biol.*, **88**, 526.
8. Waggoner,A.S. (1979) in *Methods in Enzymology,* Fleischer,S. and Packer,L. (eds.), Vol. 55, Academic Press, New York, p. 689.
9. Akerman,K.E.O. and Wikstrom,M.K.F. (1976) *FEBS Lett.*, **68**, 191.
10. Bashford,C.L., Alder,G.M., Gray,M.A., Micklem,K.J., Taylor,C.C., Turek,P.J. and Pasternak,C.A. (1985) *J. Cell. Physiol.*, **123**, 326.
11. Chance,B. and Williams,G.R. (1955) *J. Biol. Chem.*, **217**, 395.
12. Hempel,F.G., Jobsis,F.F., LaManna,J.L., Rosenthal,M.R. and Saltzman,H.A. (1977) *J. Appl. Physiol.*, **43**, 873.
13. Bashford,C.L., Barlow,C.H., Chance,B., Haselgrove,J.C. and Sorge,J. (1982) *Am. J. Physiol.*, **242**, C265.
14. Estabrook,R.W. (1956) *J. Biol. Chem.*, **223**, 781.
15. Kerr,S.E. (1935) *J. Biol. Chem.*, **110**, 625.
16. Bashford,C.L. (1981) *Biosci. Rep.*, **1**, 183.
17. Bashford,C.L., Chance,B. and Prince,R.C. (1979) *Biochim. Biophys. Acta,* **545**, 46.
18. Geisow,M.J. (1984) *Exp. Cell Res.*, **150**, 36.
19. Geisow,M.J. (1984) *Exp. Cell Res.*, **150**, 29.
20. Rottenberg,H. (1979) In *Methods in Enzymology,* Fleischer,S. and Packer,L. (eds.) Vol. 55, Academic Press, New York, p.547.
21. Tsien,R.Y., Pozzan,T. and Rink,T.J. (1982) *J. Cell Biol.*, **94**, 325.
22. Rogers,J, Hesketh,T.R., Smith,G.A. and Metcalfe,J.C. (1983) *J. Biol. Chem.*, **258**, 5994.
23. Junge,W.W., Auslander,W., McGeer,A.J. and Runge,T. (1979) *Biochim. Biophys. Acta,* **546**, 121.
24. Bashford,C.L., Alder,G.M., Micklem,K.J. and Pasternak,C.A. (1983) *Biosci. Rep.*, **3**, 631.
25. Bashford,C.L., Foster,K.A., Micklem,K.J. and Pasternak,C.A. (1981) *Biochem. Soc. Trans.*, **9**, 80.
26. Sims,P.J., Waggoner,A.S., Wang,C.-H. and Hoffman,J.F. (1974) *Biochemistry,* **13**, 3315.
27. Bashford,C.L., Barlow,C.H., Chance,B. and Haselgrove,J. (1980) *FEBS Lett.*, **113**, 78.
28. Smith,J.C., Russ,P., Cooperman,B.S. and Chance,B. (1978) *Biochemistry,* **15**, 5094.

CHAPTER 6

Stopped-flow Spectrophotometric Techniques

JOHN F. ECCLESTON

1. INTRODUCTION

1.1 Rapid Reactions

Conventional spectrophotometers and fluorimeters cannot usually follow reactions which occur before 3–5 sec after mixing reagents, this being the time it takes to mix the reagents, close the lid of the sample compartment and activate the instrument (which may also have a rather long response time). Many important biological processes occur on a time scale much shorter than this. For example, enzymes turn over substrates at rates of between 0.1 and 1000 sec^{-1}, muscle fibres can generate force in less than 10 msec, ions are moved across membranes by transport ATPases at approximately 50 sec^{-1} and, in the biosynthesis of proteins, 20 amino acids per second are added to the growing peptide chain. To gain information about the mechanism of such processes and their regulation, it is necessary to be able to make measurements on a millisecond time scale. The sensitivity and selectivity of spectrophotometry and fluorimetry combined with rapid mixing methods allows this type of measurement to be made.

This chapter describes the technique of stopped-flow spectrophotometry and includes a general description of the apparatus, procedures for testing the performance of the apparatus and the principles used in the design and the analysis of the experiments. It is assumed that the reader has access to either a commercial or a home built apparatus or a stopped-flow adaptation of a conventional spectrophotometer. More detailed information about the material covered here may be found in the books by Gutfreund (1) and by Hiromi (2) which also cover other methods for studying rapid reactions such as temperature- and pressure-jump measurements.

1.2 Historical Development and Principle of Operation

The principle of stopped-flow spectrophotometry is best explained by a description of the instrument constructed by Hartridge and Roughton in 1923 to measure the rate of binding of carbon monoxide to haemoglobin. Solutions of each reactant were continuously driven through a mixing chamber at constant pressure out into an observation tube. Since the flow of both of the solutions through the mixer is constant, the age of the mixed solution at any point in the observation tube is obtained by dividing the volume between the mixer and that point by the flow rate. The binding of carbon monoxide to haemoglobin is exothermic and so a small temperature change as well as a change in light absorption accompanies the reaction. A thermocouple or a Hartridge reversion spectroscope placed at different positions along the observation tube allowed measurements to be made on reactions with half-times as short as 1 msec. Unfortunately, this continuous flow method, which has excellent time resolution and does not need rapidly responding detectors, requires large volumes of solution; this is a major disadvantage for most biological systems. The availability of rapid electronic methods for light detection led to the development of the stopped-flow method by Chance in 1940. In this method, the flow of the two solutions is suddenly stopped; the mixed solution in the observation chamber is a few milliseconds old and a physical parameter, usually spectrophotometric, is then followed with time. Much smaller volumes are needed for this method compared with continuous flow methods, an important advantage for precious biological samples.

It is surprising that in the four decades since the first stopped-flow instrument was used, little has been gained in time resolution. In fact, a predecessor of the Hartridge and Roughton instrument was one described by Raschig in 1905 in which a gaseous reaction was studied at times down to 25 msec. The availability of photomultipliers has, however, increased the sensitivity of detection and the development of microcomputers has revolutionised the analysis of data.

2. ESSENTIAL FEATURES OF THE INSTRUMENT

A stopped-flow spectrophotometer is essentially a conventional spectrophotometer (see Chapter 1) with the addition of a system for rapid mixing — the 'stopped-flow block'. This component will be dealt with first.

2.1 Mixing and Observation System

A diagrammatic representation of the stopped-flow block is shown in *Figure 1a*. The two syringes (B) contain the reactant solutions. Flow is initiated by movement of the

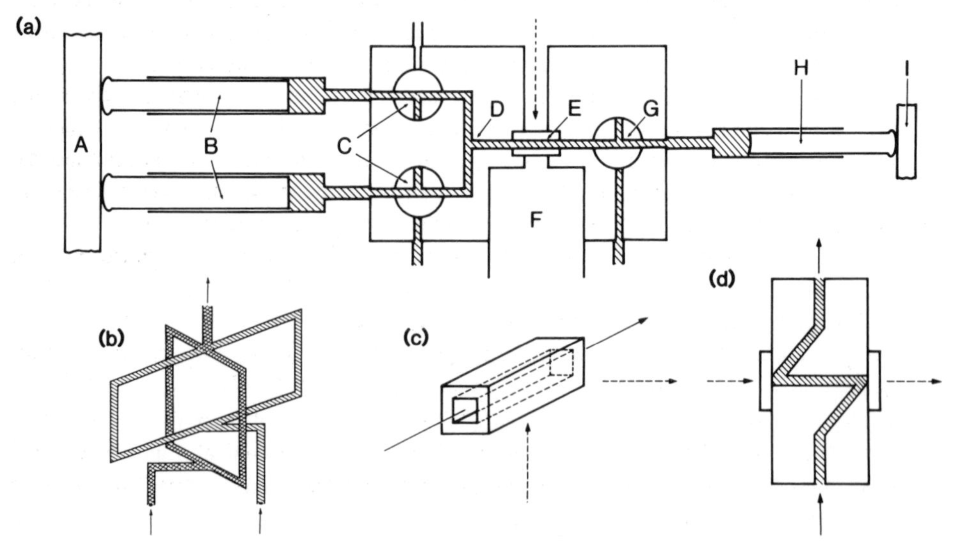

Figure 1. Diagrammatic representation of a stopped-flow instrument. **(a)** Plan view of an instrument stopped by the back syringe. A, drive; B, reactant syringes; C, filling valves; D, mixer; E, observation cell; F, photomultiplier; G, emptying valve; H, back syringe; I, stopping bar. **(b)** A four-jet mixer of the type used for most of the experiments shown in this chapter. **(c)** Fluorescence observation cell with straight-through flow. **(d)** Absorption observation cell of 'Z' shape configuration. Liquid flow is represented by solid arrows and the light path by dashed arrows.

drive plate (A). Good results can be obtained by an experienced operator if this is done manually. However, on many instruments, the drive is operated by a pneumatic or hydraulic device and this is probably better for inexperienced operators and for the collection of a series of traces for computer averaging.

The solutions from the two syringes then flow into the mixing chamber (D). In its simplest form, this can simply be a 'T' shape connection. More complex mixers are usually used in order to increase the mixing efficiency of the device. This becomes important if solutions of high viscosity or with very different optical properties are used. A four-jet mixer of the type used for most of the experiments illustrating this chapter is shown in *Figure 1b*. In this the solution from each syringe is divided into two streams. The four jets of solution are then mixed tangentially before flowing into the observation cell. Mixers with more than four jets and with more complex systems of mixing have been described.

The observation cell (E) is constructed either of glass or quartz depending on the wavelength of the light being used. For fluorescence measurements the cell is usually of a straight flow-through design with a square cross-section of $1-4$ mm^2 (*Figure 1c*) in which the incident light passes into one surface and emitted light is monitored at 90°. This cell can also be used for absorption measurements by monitoring light at 180° to the incident light. This will only measure absorption over a short path length ($1-2$ mm) so a 1 cm path length 'Z' configuration cell is usually used for these measurements (*Figure 1d*). The disadvantage of the 'Z' configuration is that the dead time (see below) is longer than that for a straight-through cell.

The solution from the observation cell then flows into the back syringe (H). In the instrument shown in *Figure 1a*, the back syringe acts as a stopping device by hitting a metal block (I). In some instruments, flow is arrested by stopping the drive syringes. The former system has the advantage of being less prone to cavitation effects (see Section 3.3) but has more stringent requirements for leakproof plumbing.

The system also contains a trigger switch which initiates the recording of the signal with time. The position of the trigger switch is usually adjusted so that the signal is observed before flow stops. During steady flow the signal remains constant and the reaction trace follows after flow stops. The advent of computerised data collection allows the signal to be monitored for a specific time prior to the trigger.

The instrument may contain a type of flow monitor which converts the linear position of the syringes to a voltage. This enables the exact time when flow stops to be correlated with the signal and also gives a record of the rate of flow.

Most instruments are designed for syringes of equal volume. It is important to remember that the concentration of each reactant in the mixing chamber is half that of the concentration in the syringe. When reporting results, it should be clearly stated whether concentrations refer to the mixing chamber or to the syringe. Some specialised purposes may require syringes of differing volumes; the dilution on mixing then needs to be calculated.

2.2 Light Source

For measurements in the visible region of the spectrum (>350 nm), a quartz halide lamp is most suitable. For the u.v. region, a xenon or xenon$-$mercury arc lamp is usually used although instruments utilising a deuterium lamp have been described. The xenon–mercury lamp gives higher intensities at the mercury lines if these can be used for the system under study. Although the arc lamps can be used in the visible region, the better stability of the quartz halide lamp makes it more suitable.

Monochromatic light is obtained by focusing the light from the lamp onto the inlet slit of a monochromator. The light from the outlet slit is then formed into a collimated beam and passed into the observation cell. Quartz light guides are now commonly used for transmitting light to the observation cell. This has the advantage of allowing the light source to be physically isolated from the mixing block and so reduces mechanical noise on the lamp source and monochromator. They also reduce the constraints on the geometry of the construction. If frequent changes of wavelength are not required monochromatic light can also be obtained by the use of band-pass or interference filters.

2.3 Signal Detection

Light from the observation cell is detected by a photomultiplier. For fluorescence measurements, filters are required to resolve the fluorescence emission from scattered excitation light. Either band-pass or cut-off type filters are suitable. The characteristics of glass filters from a wide range of manufacturers have been documented by Dobrowolski *et al.* (3) and Kodak produce a booklet showing the transmission spectra of their gelatin filters (4). The aim of filter selection is to maximise the signal while keeping scattered light to a minimum. This is illustrated in *Figure 2*.

The output of the photomultiplier is included in a circuit which allows for the incorporation of signal offset, signal amplification and noise reduction (by imposing a defined

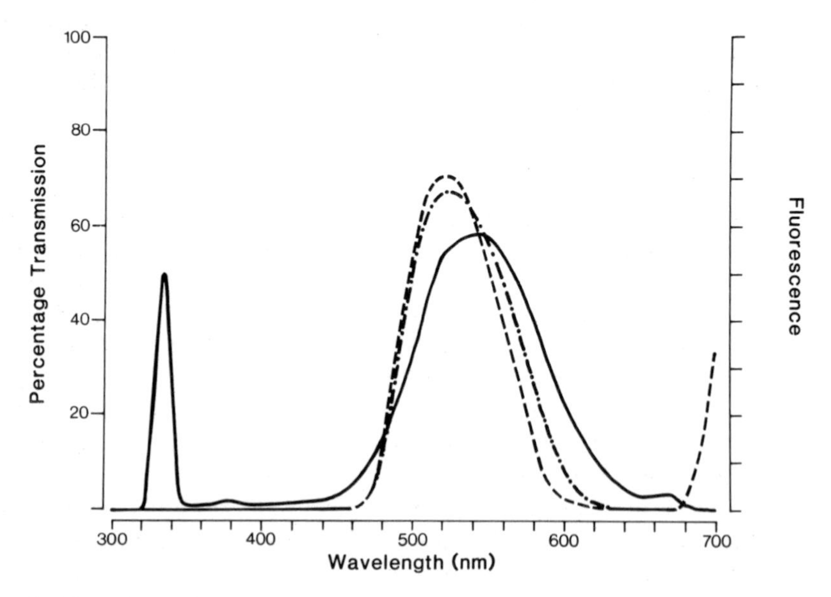

Figure 2. Filter selection for observation of dansyl fluorescence. The solid line shows the uncorrected fluorescence emission spectrum of dansylglycine excited at 335 nm. The emission maximum is at 550 nm. The other lines are the transmission spectra of Wratten 55 (------) and 57 (-·-·-·-·-) gelatin filters. Both filters pass a reasonable amount of the emitted light and completely block scattered excited light (335 nm), although the 55 filter is just starting to pass the second harmonic scattered light at 670 nm. For scattering solutions this could reduce the amplitude of the signal.

time constant). If the instrument is used in the transmission mode, the signal should be converted from transmission to absorbance. This can be achieved electronically or by processing the data, manually or by computer, after it is collected. However, if the overall change in transmission is less than 20%, it is reasonable to assume that changes in concentration are proportional to the changes in transmission.

The record of the reaction may be displayed on an oscilloscope and photographed with a 35 mm camera which gives records in a form for easy filing and also allows the record to be enlarged and traced onto graph paper for subsequent analysis. Polaroid cameras give an immediate, permanent record but are expensive. Photographs are ideal for preliminary experiments on a new system when the effect of reaction concentrations and wavelength are being explored. A strip of 35 mm film allows a rapid comparison of traces and also allows a preliminary analysis of rate constants. However, for accurate determination of rate constants it is best to send the signal directly to a transient recorder where it is stored in digital form suitable for subsequent computer-based analysis.

3. THE STOPPED-FLOW EXPERIMENT

3.1 Choice of Signal

3.1.1 *Intrinsic Probes*

The first requirement for a successful stopped-flow experiment is the presence of a suitable absorption or fluorescence change accompanying the process being studied. Nature has been kind to the spectroscopist in providing tryptophan residues whose fluorescence often characterises different states of the protein; NADH which loses its absorption band at 340 nm (and its accompanying fluorescence emission at 450 nm) when oxidised to NAD^+; and metalloproteins which display a wide range of spectra in their different oxidation states. All of these chromophores have been extensively used in stopped-flow studies.

3.1.2 *Extrinsic Probes*

If an intrinsic chromophore is not available for the system, it is often possible to introduce one, and many compounds exist for labelling proteins with suitable chromophores or fluorophores. Remember that few of these reagents are truly specific and may react with more than one amino acid residue of the protein. This may not be critical for stopped-flow experiments but it is good policy to characterise the derivatised material. Details of labelling reagents together with references have been documented by Haugland (5). Alternatively, natural small molecules can be modified to give suitable spectroscopic properties, for example many analogues of ATP with a wide range of properties have been synthesised and can be used for stopped-flow studies (6).

If an extrinsic chromophore has been introduced, its effects on the system should be investigated by as many methods as possible. The probe may dramatically alter the rate and equilibrium constants of a particular process, but may still promote the biological action of interest. For example formycin 5′-triphosphate, a fluorescent analogue of ATP, causes muscle fibres to contract and the chromophoric analogues of guanosine nucleotides derived from 2-amino-6-mercaptopurine riboside are active in the elongation cycle of protein biosynthesis. The quantitative effects of these analogues differ from those of natural nucleotides but the analogues still may provide qualitative information about the system; the rationale being that although the individual rate constants of the elementary processes of the mechanism may differ, it is likely that the overall mechanism of the process is the same with both natural compound and analogue.

Ideal extrinsic chromophores do not perturb the system but give large signal changes. Thus, for example, actin labelled at Cys 374 with N-(1-pyrenyl) iodoacetamide interacts with a subfragment of myosin with a concomitant 70% quenching of the pyrene fluorescence (7). The original paper gives further details of the experiments and is a good example of a detailed investigation into the effects of the introduction of the fluorophore into a system.

Although some uncertainty is introduced by the use of modified proteins or ligand/substrate analogues, they can be used to give information about the physiological compounds by the use of competition and displacement reactions. These are discussed in Sections 5.5 and 5.6.

3.1.3 *Quenching of Fluorescence*

If the absorption spectrum of a molecule overlaps the fluorescence emission spectrum of a second molecule, quenching of fluorescence may occur by dipole–dipole interaction. This is the basis of a method for distance measurements within macromolecular complexes since the extent of quenching is dependent on the sixth power of the distance between the molecules. The effect can also be used as a signal for stopped-flow experiments. In principle, any ligand with absorption in the region of 310–360 nm may cause quenching of tryptophan fluorescence and thus monitor protein–ligand interaction. It should be noted, however, that in some systems the quenching of protein fluorescence may not be linear with respect to complex formation, and if this is not taken into account incorrect values of rate constants will be obtained. This effect has been well documented in studies on the binding of NADH to the tetrameric enzyme lactate dehydrogenase (8).

Alternatively, fluorescence may be quenched by collisional mechanisms. A small molecule such as acrylamide or iodide colliding with an excited state molecule may deactivate it and therefore cause quenching of fluorescence. This gives a method for measuring the accessibility of tryptophan residues in proteins to solvent. However, it can also be used to enhance fluorescence changes of a ligand on binding to a protein if the bound ligand is inaccessible to the quenching agent, i.e. the bound ligand fluoresces more strongly than the free ligand. This technique has been used to study the binding of etheno-ATP to myosin in the presence of acrylamide (9).

3.1.4 *Linked Assays*

During a biological process, a particular species may be released by the system. Even if no spectral change is associated with the process, it may be observed by linking it to another process which does exhibit a spectral change. This is the basis of classical linked-enzyme assays described in Chapter 3 and the primary requirement is that the secondary reaction is so much faster than the one being studied that it does not affect the observed rate. This condition makes measurements difficult on the typical time scale of stopped-flow experiments, although some measurements have been made using this technique. For example, in the presence of NAD^+, D-glyceraldehyde-3-phosphate, ADP, D-glyceraldehyde-3-phosphate dehydrogenase (GPDH) and phosphoglycerate

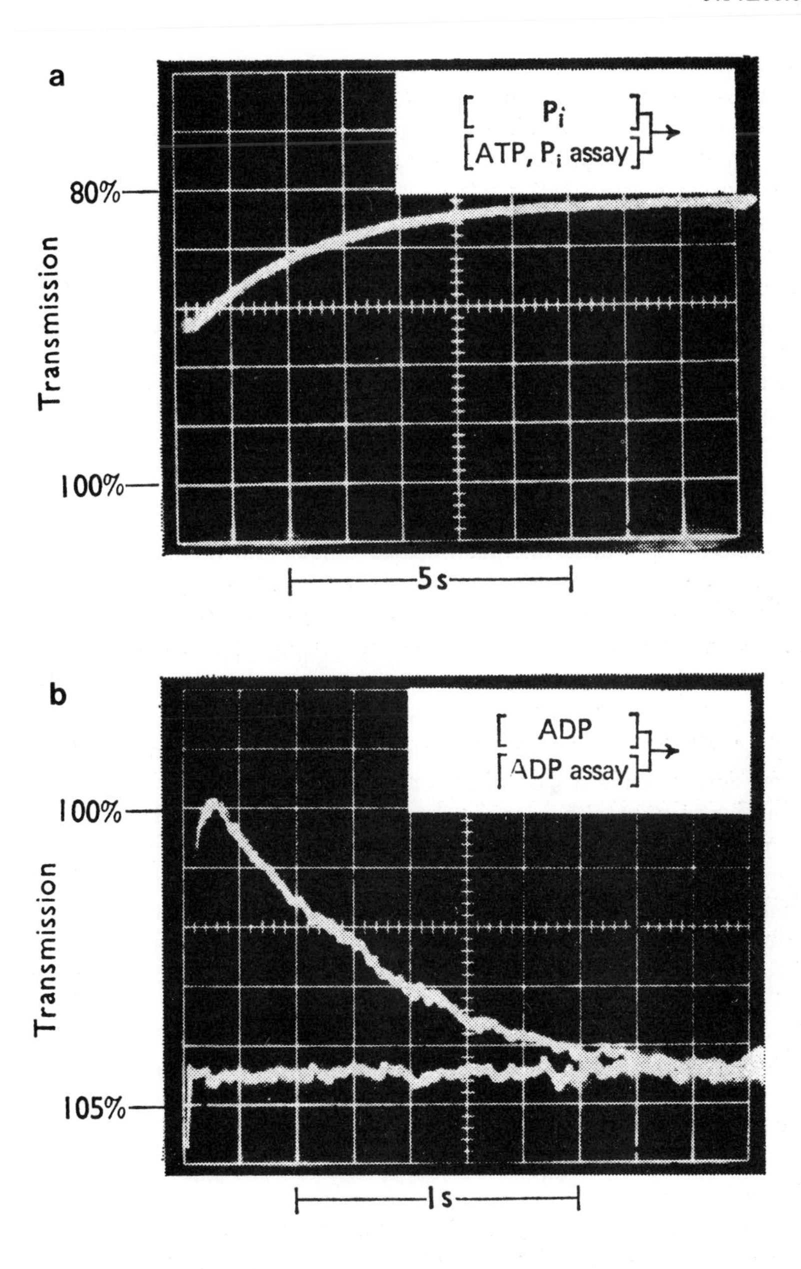

Figure 3. Linked assay measurements. **(a)** Assay of P_i measured by NADH production in a linked assay system. One syringe contained 10 μM P_i and the other contained 0.68 mM D-glyceraldehyde 3-phosphate, 1.6 mM NAD^+, 2.8 mM $MgCl_2$, 0.34 mM ADP, 0.38 mg/ml GPDH and 2.4 μg/ml PGK (reaction chamber concentrations). Both syringes also contained 80 mM triethanolamine HCl, pH 8.0. **(b)** Assay of ADP measured by NADH disappearance in a linked assay system. One syringe contained 5 μM ADP and the other 0.14 mM NADH, 2.8 mM PEP, 37 mM Tris-HCl, pH 8.0, 1 mg/ml PK and 0.5 mg/ml LDH. Both syringes also contained 3.7 mM $MgCl_2$, 37 mM KCl and 37 mM triethanolamine HCl, pH 8.0. The horizontal trace was recorded a few seconds after the end of reaction. All records were made by following transmission at 340 nm in a 1 cm path length cell. (Ref. 10.)

kinase (PGK), inorganic phosphate (P_i) released in a reaction is linked to the conversion of NAD^+ to NADH by GPDH and so can be monitored by observation of NADH absorption or fluorescence. PGK and ADP are present to remove the product of the first reaction and thus pull the equilibrium in favour of NADH. The reactions are:

$$G3P + P_i \xrightarrow[NAD^+ \quad NADH]{GPDH} 1,3\ DPG \xrightarrow[ADP \quad ATP]{PGK} 3\ PG$$

A similar stopped-flow assay for measuring the release of ADP has also been described by linking it to the pyruvate kinase and lactate dehydrogenase systems:

$$PEP \xrightarrow[ADP \quad ATP]{PK} pyruvate \xrightarrow[NADH \quad NAD]{LDH} lactate$$

However, both of these reactions are slow, the former having a half-time of about 2 sec and the latter 0.5 sec under the published conditions (10) (*Figure 3*).

3.1.5 *Indicators*

Experiments using indicators are not usually thought of as linked assays although the principle is the same; for example the concentration of H^+ is linked to the pH indicator so that changes in $[H^+]$ are accompanied by absorption or fluorescence changes. These are both rapid (ionic interactions occur at about 10^{10} M^{-1} sec^{-1}) and reversible so that either the uptake or the release of protons can be measured.

pH indicators register changes in pH most sensitively when pH is equal to the p*K* of the indicator so the choice indicator is governed by the pH of the experimental conditions. A list of commonly used indicators with their properties is given in *Table 1*. The amplitude of the pH change for a given process in which protons are taken up or released decreases with increasing buffering capacity of the solution. The latter is comprised both of the buffer itself and any components of the system, such as proteins and the indicator itself, which can act as a buffer.

A good practical aim is to work at a buffer concentration low enough to give a reasonable signal change but high enough to maintain the pH of the solution reliably at the

Table 1. Properties of pH indicators.

Compound	*pK*	*mol. wt*	*λ max (nm)*	10^{-3} ϵ M^{-1} cm^{-1}
Thymol blue	1.4	466	550	33
Benzyl orange	2.6	405	505	56
Bromo Cl phenol blue	4.0	569	590	72
Bromo cresol green	4.7	698	610	42
Chloro phenol red	5.9	423	580	44
Bromothymol blue	6.9	624	620	36
p-Nitrophenol	7.0	139	400	13
Phenol red	7.9	354	560	54
Cresol red	8.1	382	525	53
Thymol blue	9.0	466	600	30
Phenolphthalein	9.6	318	552	33

required value. Concentrations in the range 0.2–1 mM are usually suitable. Under these conditions the solutions tend to decrease in pH due to the uptake of atmospheric CO_2. A good way of obtaining both solutions at the optimal conditions is to dialyse the protein solution against these weakly buffered solutions using a pH stat to maintain the pH at a value slightly below the pH of the final reaction. After the addition of a suitable aliquot of indicator to both solutions to give an absorbance in the range of 0.5–1.0, both solutions are adjusted to the final pH in a pH stat, titrating in 5 mM KOH. The stopped-flow syringes are then rapidly loaded, if possible, anaerobically.

An artifact can arise in these experiments in addition to those described in Section 3.3. It is important to check that the indicator does not bind to any components of the system which may then give spectral changes during the reaction which are not associated with changes of pH. The best check on this problem is to compare the spectrum and pK of the indicator in the presence and absence of protein in buffered solutions, using a scanning spectrophotometer (6). A good example of pH indicator experiments is that of Finlayson and Taylor (11).

Indicators are also available for monitoring ions other than protons: for example, Quin 2 is a fluorescent compound which chelates Ca^{2+} (12). The fluorescence signal (excitation 339 mm, emission 492 mm) increases about 5-fold on the formation of a 1:1 complex with Ca^{2+} with a K_d in the region of 100 nM. Stopped-flow experiments in which a Ca^{2+}–Quin 2 complex is mixed with EDTA show the dissociation rate of the complex is about 60 sec^{-1}. Since the establishment of equilibrium is very fast, this reagent is suitable for rapid kinetic measurements of changes in Ca^{2+} concentration (13). These are illustrated in *Figure 4*.

3.1.6 *Overall Considerations*

Very often experimental design will be limited by the nature of the signal that can be observed and the concentration of reactants available. However, if a choice of signals is available, it is very helpful to maximise the signal by the appropriate choice of conditions, particularly if a wide concentration range is to be used. In a reaction of the type

$$A + B \rightleftharpoons AB$$

where B is in large excess over A, then it is much more favourable if a signal corresponding to the loss of A or formation of AB is observed, rather than one associated with the loss of B. This is illustrated in *Table 2* where the parameters for a hypothetical reversible second-order reaction are given (see Section 5.3 for details of rate constants). It can be seen that under the conditions given, the reaction could be followed over a wide range of concentrations of B if AB is observed. However, if the loss of B is observed, the signal will become too small to measure at higher concentrations of B.

As a practical example, the formation and decay of an intermediate when myosin subfragment 1 interacts with ATP or ATP analogues has been investigated by several different methods (*Figure 5*) (6,14). In the first instance the absorption change when 6-mercaptopurine riboside 5′-triphosphate (thioITP) interacts with subfragment 1 was monitored using concentrations up to 100 μM; at higher concentrations the solution is too opaque for reliable measurements to be made. The same analogue quenches the fluorescence of subfragment 1 and gives a relatively large signal to monitor. However,

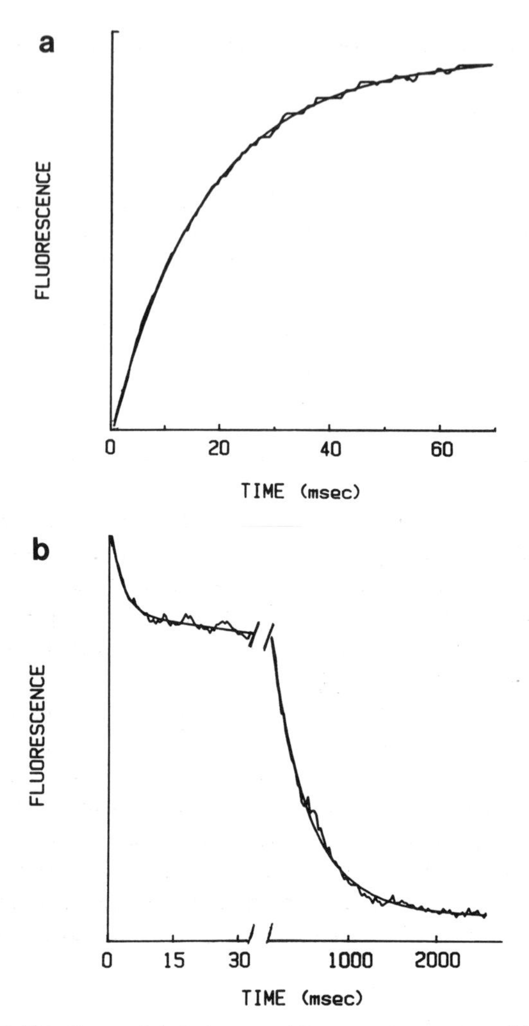

Figure 4. Use of the Ca^{2+} indicator, Quin 2, in stopped-flow experiments. **(a)** Ca^{2+} dissociation from the Ca^{2+}–Quin 2 complex. One syringe contained 25 μM Quin 2, 50 μM Ca^{2+} and the other contained 5 mM EDTA (reaction chamber concentrations). The line drawn through the data represents a fit to the data and k_{obs} = 57 sec^{-1}. **(b)** Ca^{2+} dissociation from a Ca^{2+}–calmodulin complex. One syringe contained 6.25 μM calmodulin, 62.5 μM Ca^{2+} and the other contained 100 μM Quin 2. The line drawn through the data represents a fit to the data with a fast process of 356 sec^{-1} and a slow process of 2.2 sec^{-1}. Excitation of Quin 2 fluorescence was at 366 nm and emission above 475 nm was measured using a Schott GG-475 filter. The two processes observed represent dissociation of Ca^{2+} from two binding sites on calmodulin which have different affinities for Ca^{2+}. After correction for the dead time of the instrument, the amplitudes of the two processes are equal. (Ref. 13.)

inner filter effects (due to the high absorbance of the analogue at both the excitation and emission wavelengths of tryptophan; see Chapter 1) again prevent the use of concentrations much above 100 μM. The fluorescent analogue, formycin 5′-triphosphate, increases its emission by 100% on binding to subfragment 1 and so gives particularly good signals. However, at high concentrations the background formycin fluorescence seriously masks the fluorescence changes due to binding. ATP itself increases the intrin-

Table 2. Calculation of observed rate constants and amplitude of signal change for a reversible bi-molecular reaction.

For the scheme $A + B \underset{k_{-1}}{\overset{k_1}{\rightleftharpoons}} AB$

where $[A_o] = 5\ \mu M$, $k_1 = 5 \times 10^5\ M^{-1}\ sec^{-1}$, $k_{-1} = 0.5\ sec^{-1}$

$[B_o]$ (μM)	k_{obs} (sec^{-1})	$t_{1/2}$ *(msec)*	*[AB] at equilibrium* (μM)	*% A bound at equilibrium*	*% B bound at equilibrium*
10	15	46	2.2	44	22
20	20	34	3.1	63	16
30	25	28	3.6	72	12
40	30	23	3.9	78	10
50	35	20	4.1	82	8

sic protein fluorescence by 20% when it binds to subfragment 1. Since ATP has little absorption above 290 nm, no inner filter effects occur and so its interaction with subfragment 1 can be studied over a very wide range of concentrations. It is fortunate in this case that the physiological nucleotide gives the best signal for monitoring the concentration dependence of the reaction, although the larger amplitude of the other signals are useful for certain studies.

It is preferable to investigate the spectral change to be observed by conventional methods (see Chapters 1, 2) before performing stopped-flow experiments. This is not always possible, as in the case of experiments designed to observe the transient appearance of an intermediate when the reactant and final product have the same spectral properties, or in reactions in which the transient disappearance of a reactant or transient appearance of a product before the process reaches a steady-state are to be monitored. When the spectral changes can be characterised by conventional methods selection of the optimal wavelengths can be made for absorption and fluorescence measurements as described in Section 2.3.

3.2 Collection of Data

Once the operating conditions have been chosen the stopped-flow spectrophotometer must be set up to record the reaction with optimal sensitivity. In general, it is best optically to balance the instrument on the end point of the reaction:

(i) Load the two reactants into syringes and equalise the positions of the syringes.

(ii) Make several pushes to ensure that the observation cell is filled with the mixed reagents.

(iii) Close the valves to prevent back-diffusion of this solution into the syringes containing unmixed reagents.

(iv) Apply a voltage to the photomultiplier so that a suitable signal voltage is obtained (usually between 1 and 6 V). At photomultiplier voltages less than 300 V the response is not linear with light input and at high voltages the signal/noise ratio decreases. Generally voltages between 400 and 700 V will provide adequate signals.

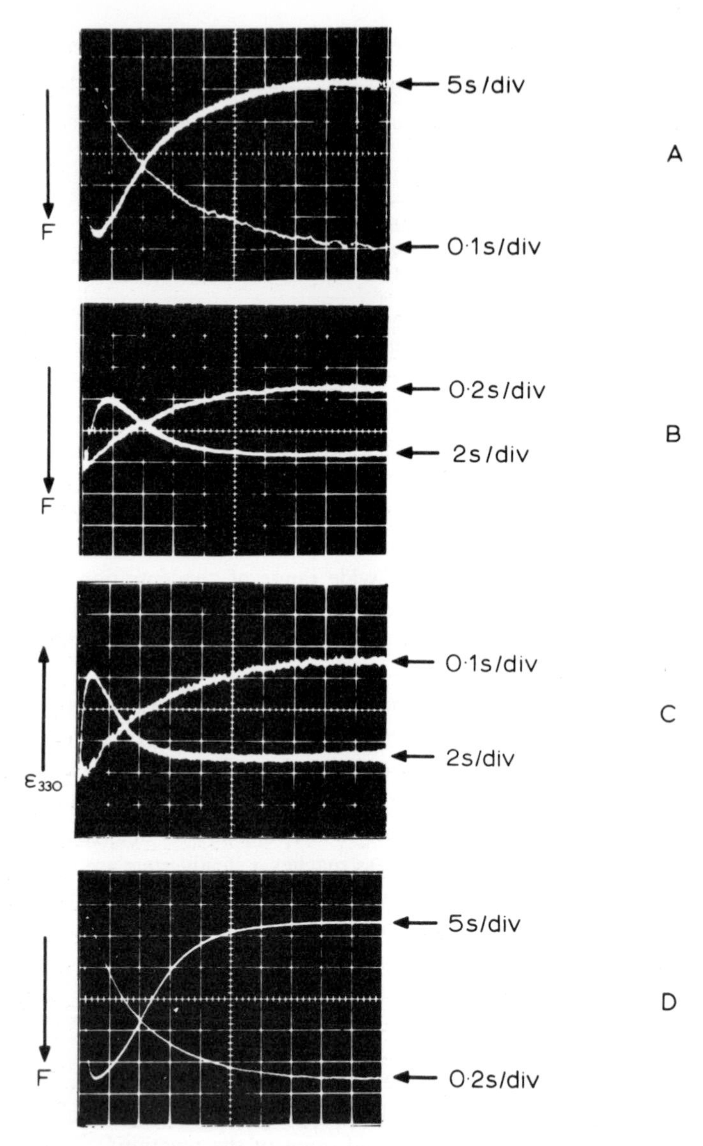

Figure 5. Records of stopped-flow experiments on the reactions of ATP and ATP analogues with the active site of myosin. These experiments were carried out under 'single turnover' conditions (active site concentration in excess of substrate). The trace on the faster time scale shows the rate which controls the enzyme–product complex formation at the particular concentration of reactants. The trace on the slower time scale represents the rate of a step which controls product dissociation. The differences in rates obtained from the different signals can be used to characterise the steps of the reaction of myosin with ATP. All four experiments were carried out in 50 mM Tris buffer, pH 8.0, 0.1 M KCl and 5 mM $MgCl_2$ and concentrations are those in the reaction chamber. **Trace A** represents a record of protein fluorescence enhancement when a solution of 3 μM ATP was mixed with a solution of 3.6 μM subfragment 1 of myosin. **Trace B** represents a record of protein fluorescence quenching when a solution of 3 μM thioITP was mixed with a solution of 5 μM subfragment 1 of myosin. **Trace C** represents a record of change in extinction at 330 nm due to change in nucleotide spectrum when a solution of 10 μM thioITP was mixed with a solution of 15 μM subfragment 1. **Trace D** represents a record of the change in nucleotide fluorescence when a solution of 5 μM formycin 5′-triphosphate was mixed with a solution of 10 μM subfragment 1 of myosin. (Ref. 14.)

In order to measure transient species which are followed by a steady-state r
the instrument must be balanced on the start of the reaction rather than the e.

(i) Fill one syringe with buffer and the other with one of the reactants (which reac
is chosen depends on the signal being observed).
(ii) Follow steps (ii) to (iv) above.
(iii) Replace the buffer by the second reactant and collect data.

This type of measurement tends to be difficult when an oscilloscope is used for recording the signal, since any drift of the baseline causes the start of the reaction to be lost from the screen. However, if the data are collected in digital form, the signal can be plotted on a suitable scale after collection of data. Data obtained from an oscilloscope trace can at best be obtained with a 6-bit precision (1 in 64) while digital recorders provide 8- or 12-bit precision which allows the scale to be expanded after the record is taken.

It remains to choose a time constant for the signal detection system (to reduce noise); usually the half-time of the detection system will be less than 20% (one fifth) that of the process under study. The instrument is now ready to collect data.

3.3 Artefacts

Many artefacts can occur in stopped-flow experiments; some of them giving rise to apparently perfect reactions! They at times simulate exponential processes, lag phases and step increases. A common problem is the presence of an air bubble in the observation cell and the stopped-flow trace is a record of the movement of this with time. Air bubbles may form as a result of cavitation which is the formation of a stream of fine air bubbles in the observation cell due to the wide variation in hydrostatic pressure at different parts of the fluid. Such bubbles decrease the transmission which subsequently increases as they re-dissolve or rise out of the light path. Effects of air bubbles can be reduced by using degassed solutions. Inefficient mixing, incomplete equilibration of temperature, leakage from the system or large differences between the densities of the two solutions can all give rise to 'reaction traces'. Suitable control experiments can usually indicate whether the reaction observed is a true process or an artefact. Pushing two solutions of buffer against each other is the easiest such control but it is better to approximate all of the conditions of the experiment but with only one reactant. For example, when protein fluorescence is being monitored to measure ligand – protein interactions, the protein should be rapidly mixed with exactly the same solution that contains the ligand but omitting the latter reagent.

If a shift of the spectrum occurs during a reaction giving a typical difference spectrum with negative and positive peaks and an isosbestic point, then the stopped-flow experiment should be performed at all three wavelengths to ascertain that the amplitude and direction of the signal are consistent with those expected. In this instance changes observed at the isosbestic point will be particularly informative about possible artefacts.

4. TESTING THE STOPPED-FLOW INSTRUMENT

Two important measurements to be made before using an instrument for the first time are those of the mixing efficiency and the dead time.

4.1 Mixing Efficiency

It is essential that thorough mixing of the two solutions occurs before they reach the observation cell of the instrument. The purpose of the mixing chamber is to induce turbulent as opposed to laminar flow in the mixed solution. Although calculations can be made to determine the linear flow velocity at which this occurs, it is preferable to determine experimentally that this is occurring. The protonation of pH indicators provides a useful system for this purpose. Since protonation reactions are instantaneous with respect to the stopped-flow time scale, the reaction trace should provide no evidence of time-dependence. If it does, mixing is still occurring after flow has stopped and is influencing the rate of the observed reaction.

The principle of the experiment is to mix an indicator in a solution above its p*K* with a solution below its p*K*. For an absorption instrument, phenol red (p*K* 7.9) can be used. One solution contains 20 μM phenol red in 0.1 M sodium pyrophosphate, pH 9, and the other solution is 0.1 M sodium pyrophosphate, pH 6. Absorbance is monitored at 560 nm. 4-Methylumbelliferone (p*K* 7.6) can be used for fluorescence instruments. In this case one solution contains 10 μM 4-methylumbelliferone in 0.1 M sodium pyrophosphate, pH 9, and the other solution is 0.1 M sodium pyrophosphate, pH 6.0; excitation is at 359 nm and emitted light beween 420 and 540 nm is recorded (*Figure 6*).

If the final experiments contain a concentration of protein which significantly increases the viscosity and thus the mixing time of the solution, then the protein should be included in the above solutions. Remember however, that many proteins non-specifically bind indicators and this may affect the p*K* and optical properties of the indicator adversely.

4.2 Determination of Dead Time

The dead time of a stopped-flow instrument is the time which elapses between mixing and observation. For example, if the time taken for the solution to flow from the mixer to fill the observation cell is 3 msec, this is the dead time of the instrument and reactions occurring faster than this will not be observed. This precise theoretical definition of dead time, however, does not hold in practice since the points of mixing and observation are not uniquely defined. Although the two solutions come into contact with each other at a definite point, complete mixing occurs during the turbulent flow of the solution from the mixer to the observation cell. Similarly, the solution is not observed at a single point in this cell but over a finite distance and therefore at a finite range of times. Since both points in the theoretical definition of dead time are floating, the best that can be done is experimentally to determine the effective dead time of the instrument. The effective dead time is the average age of the solution in the observation chamber when flow stops. This is achieved by measuring the signal from the starting point of a reaction (i.e. that from the unmixed reactant from which the signal arises) and then monitoring the reaction and extrapolating the signal from this back to the starting point. The intersection between the reaction trace with the starting signal (at a negative time value) gives the value of the effective dead time.

It is difficult to devise a reaction giving a linear profile in the millisecond time range and a common procedure is to perform a pseudo-first order reaction (see below) and plot the log of the concentration against time to give a linear plot. Even this require-

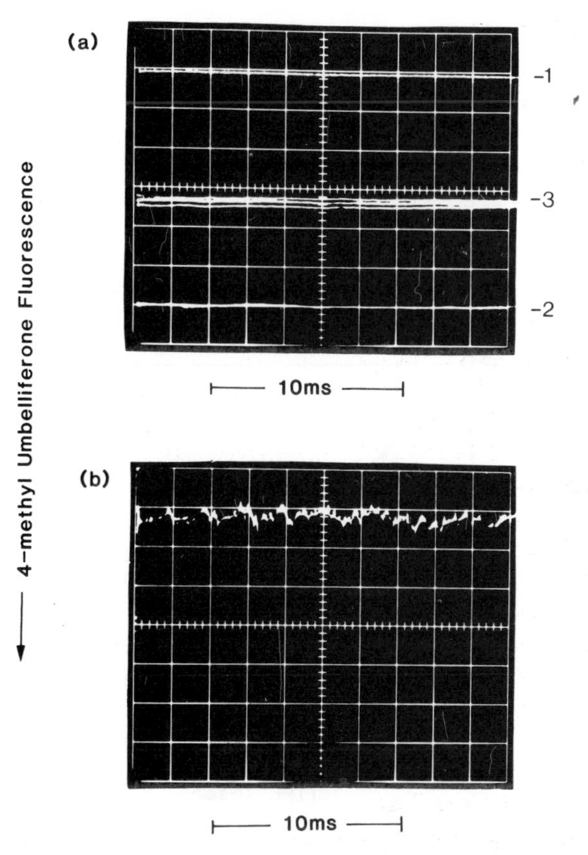

Figure 6. Test of the mixing efficiency of a fluorescence stopped-flow apparatus. Excitation was at 359 nm and the emitted light between 420 and 540 nm was recorded. **(a) Trace 1** corresponds to zero fluorescence and practically equalled the fluorescence of 9 μM 4-methylumbelliferone at pH 6.2. **Trace 2** represents the fluorescence of 9 μM 4-methylumbelliferone at pH 8.7. **Traces 3**, comprising three almost superimposed traces represent the fluorescence of 9 μM 4-methylumbelliferone at pH 7.7 and were obtained by mixing the umbelliferone in 0.1 M sodium pyrophosphate adjusted to pH 8.7 with HCl with 0.1 M sodium pyrophosphate adjusted to pH 6.2 with HCl. **(b)** An oscilloscope trace of the same reactions as **trace 3** in **a** with the ordinate amplified 20-fold so that one major division represents 1.6% of the total fluorescence change in the reaction. (Ref. 15.)

ment is not easily satisfied although two reactions are in regular use.

The first is the alkaline hydrolysis of 2,4-dinitrophenyl acetate. This compound is hydrolysed by NaOH to dinitrophenol with a second-order rate constant of approximately 60 M^{-1} sec^{-1}. Therefore, mixing 2,4-dinitrophenyl acetate with 1 M NaOH (final concentration) will give a reaction of 60 sec^{-1} ($t_{1/2}$ = 11.5 msec). The product, 2,4-dinitrophenol, occurs as an anion at alkaline conditions, and its formation can be monitored by its absorption at 360 nm. In practice, a 12.5 mM solution of 2,4-dinitrophenyl acetate (preferably recrystallised from ethanol) in isopropanol is prepared and diluted 10-fold with 5 mM HCl just prior to use. This is mixed with 2 M NaOH (syringe concentrations) in the stopped-flow spectrometer, and A_{360} recorded. Determination of dead time using this reaction is shown in *Figure 7*.

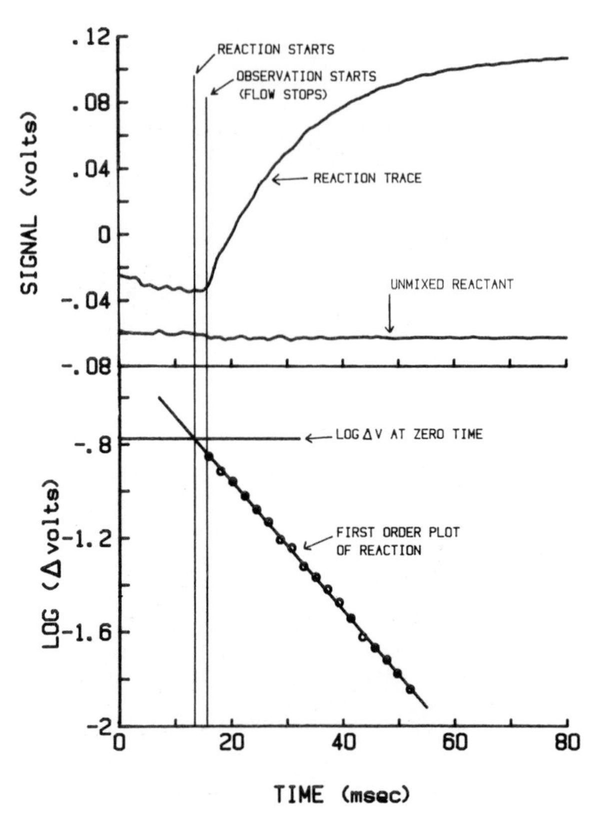

Figure 7. Determination of the dead time of a stopped-flow instrument. The reaction used was the alkaline hydrolysis of 2,4-dinitrophenyl acetate. For the reaction trace, one syringe contained 1.24 mM 2,4-dinitrophenyl acetate (prepared as a 12.5 mM solution in isopropanol) and 5 mM HCl and the other syringe contained 2 mM NaOH (syringe concentrations). Absorption was monitored through a 2 mm path length cell at 360 nm. 1 volt is equivalent to 1 A.U. The reaction was followed for a longer time (200 msec) than shown here to obtain an accurate value for the end point. This value (V_∞) was 0.109 V. For the 'unmixed reaction' trace, the 2 M NaOH was replaced by 5 mM HCl to obtain a value for V_0, the voltage at the beginning of the reaction. For both traces, the oscilloscope was used in the pre-trigger mode. The recording started 16 msec before flow stopped. A first-order plot of the reaction is shown below. Log ($V_\infty - V$) was plotted against time. Extrapolation of the straight line back to the value of log ($V_\infty - V_0$) gives the dead time of the instrument (2.4 msec). This is the time elapsing between the start of the reaction and the beginning of the observation. The slope of the first-order plot gives a value of the observed rate constant to be 63.6 sec^{-1}.

An alternative reaction is the reduction of 2,6-dichlorophenolindophenol by ascorbic acid. This has a rate constant of 1.04×10^4 M^{-1} sec^{-1} at pH 6.0 and can also be carried out under pseudo-first order conditions. One syringe contains the indophenol (A_{600} = 0.5) in 0.5 M phosphate, pH 6.0 and the other contains 40 mM ascorbic acid in 0.5 M phosphate pH 6.0. The loss of colour of the indophenol is monitored by absorption at 600 nm. Since the final concentration of ascorbic acid is 20 mM, a reaction with a half-time of 3.3 msec will be observed. This can be altered by varying the ascorbic acid concentration.

Both of the above reactions are for use in absorption instruments although if the light input to the observation cell of fluorescence instruments can be modified by 90°, they

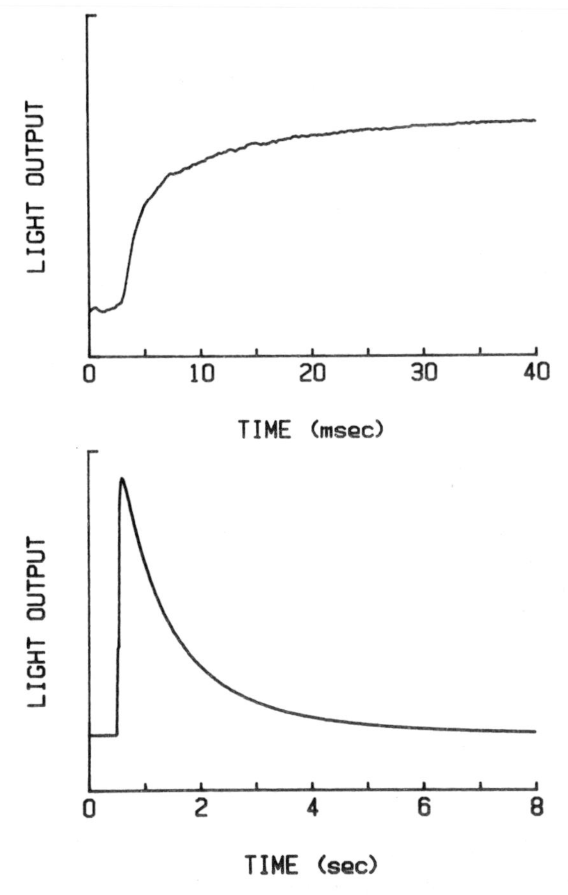

Figure 8. Stopped-flow record of the formation (upper trace) and decay (lower trace) of chemiluminescence when a solution containing 2 mM luminol, 0.45 M NaOH and 0.13 M H_2O_2 was mixed with 1.5 mM $K_3Fe(CN)_6$ in a 2 mm path length cell (syringe concentrations).

can also be used for these. The amplitude of the absorption change is large enough to be monitored using a path length of 1−2 mm.

A possible test reaction for fluorescence instruments is the binding of dansylglycine to bovine serum albumin (BSA). This reaction is accompanied by an approximately 20-fold increase in fluorescence when excitation is at 330 nm and emission is measured through a Wratten 57 filter. The observed first-order rate constant for the reaction when 5 μM BSA is mixed with 1 μM dansylglycine (reaction chamber concentrations) is 37 sec^{-1}. This is somewhat slow for accurate dead time determinations. If either reactant is increased in concentration, the reaction profile becomes biphasic, also making it unsuitable for this purpose. It is, however, a useful reaction for learning to operate a stopped-flow spectrofluorimeter since both reagents are readily available.

The chemiluminescent reaction of luminol has been suggested as a test reaction. Mixing a solution of 2 mM luminol in 0.45 M NaOH and 0.13 M H_2O_2 with 1.5 mM $K_3Fe(CN)_6$ (syringe concentrations) results in a rapid production of light followed by a slow decay (*Figure 8*). Obviously, no light source is required for this reaction, and it can be monitored in both absorption and fluorescence instruments. An advantage of this reaction is that the signal at the start of the reaction is the same as that at the end, so both the reaction profile and the starting point can be determined from one experiment. The disadvantage is that the initial rapid rise in light intensity does not appear to be a simple exponential and so only a rough estimate of the dead time can be obtained.

5. RATE EQUATIONS

In order to understand the procedures for analysing the data from stopped-flow experiments the rate equations for simple first- and second-order reactions are given below. However most biological reactions are usually much more complex and do not obey these rate laws in these simple forms. Even apparently simple reactions, such as the hydrolysis of ATP or of peptide bonds by enzymes, have been dissected into many steps by a combination of stopped-flow and other kinetic and equilibrium techniques. No general approach to the analysis of such systems can be given here but a fuller exposition can be found in the books by Gutfreund (1) or by Hiromi (2) and in other specialised texts. However, some examples of more complex reactions are described below which, while applicable to the analysis of enzyme mechanisms, have wider applications to the study of biological processes.

5.1 **First-order Reactions**

In a first-order reaction, the rate of reaction is directly proportional to the concentration of one reactant.

Consider the reaction: $A \rightarrow B$

$$\frac{-d[A]}{dt} = k[A] \qquad \text{Equation 1}$$

where k is the first-order rate constant for the reaction.

Rearranging gives

$$\frac{-d[A]}{[A]} = kdt \qquad \text{Equation 2}$$

Integration of this equation leads to

$$\ln([A_o]/[A]) = kt \qquad \text{Equation 3}$$

or

$$\ln[A_o] - \ln[A] = kt \qquad \text{Equation 4}$$

where $[A_o]$ is the initial concentration of A, and [A] is the concentration of A at time t.

Since $[B] = [A_o] - [A]$ and $[A_o] = [B_\infty]$, Equation 4 can be rewritten as:

$$\ln[B_\infty] - \ln([B_\infty]-[B]) = kt \qquad \text{Equation 5}$$

where $[B_\infty]$ is the concentration of [B] at the end of the reaction and B is the concentration of B at time t.

Two general consequences arise from these first-order rate equations. First, the time taken to complete a definite fraction of the reaction is independent of the initial concentration of reactant. For example, for the reaction to proceed to 50% completion (i.e.

when $[A_0]/[A] = 2$) then substituting for $[A_0]/[A]$ in Equation 3 gives:

$$t_{1/2} = \frac{\ln 2}{k} = \frac{0.69}{k} \qquad \text{Equation 6}$$

where $t_{1/2}$ is the time taken for half the reaction to occur. This provides a rapid method of evaluating the value of k from the half-time of a first-order reaction.

Second, the value of $[A_0]/[A]$ is a ratio of concentrations and so its value is independent of the units of concentration used. From Equations 4 and 5 it can be seen that a plot of $\ln[A]$ against t or $\ln([B_\infty]-[B])$ against t gives a straight line of slope k (units, sec^{-1}). If logarithms to the base 10 rather than natural logarithms are plotted, the slope is $k/2.3$. Similarly a plot of any parameter, for example absorption or fluorescence, which is directly proportional to [A] or $[B_\infty]-[B]$ gives the same result. This is illustrated in *Figure 7*. (The example shown in *Figure 7* is not first-order but the analysis is valid for reasons described below.) From these plots, values of k can be calculated. For the reaction to be truly first-order, the first-order plot described above should be a straight line for at least the first 95% of the reaction, although limitations of signal/noise ratio often prevent this being achieved in practice.

An alternative method of calculating the rate constant of a first-order reaction is based on the algebraic procedure of Guggenheim (1). This procedure has the advantage of obtaining the rate constant when neither the initial concentration of reactant nor the end point of the reaction are known. Absorbance (or fluorescence) readings are taken at equal time intervals (Δt) and a plot of $\ln(A_t - A_{t+\Delta t})$ against time (t) provides a straight line of slope k. If the reaction is followed by a steady-state rate (or is superimposed on a sloping baseline) the extended Guggenheim procedure can be used: sets of three concentration measurements are taken at equal time intervals and a plot of $\ln(A_t + A_{t+2\Delta t} - A_{t+\Delta t})$ against time gives a straight line of slope k.

5.2 Second-order Reactions

Since the essential feature of a stopped-flow experiment is the initiation of a reaction by the rapid mixing of two reactants, true first-order reactions of the type described above are not usually observed. For the preliminary analysis of typical stopped-flow reactions, we need to consider the reaction

$$A + B \rightarrow AB$$

Studies of such reactions are best made at one of two possible conditions of concentration.

5.2.1 *Initial Concentrations of Reactants are Equal*

When the concentrations of both reactants are made equal, the rate equation is

$$\frac{d[AB]}{dt} = k\,[A]\,[B] = k[A]^2 \qquad \text{Equation 7}$$

Integration gives

$$\frac{1}{[A]} - \frac{1}{[A_0]} = kt \qquad \text{Equation 8}$$

where $[A_0]$ is the initial concentration of A and [A] is the concentration at time t.

Table 3. Effect of reactant concentration on the observed first-order rate constant of pseudo-first order reactions.

The reaction A + B → C was simulated using a value for the second-order rate constant of 2×10^7 M^{-1} sec^{-1}, and 1 μM A. B was varied between 1 μM and 100 μM. The reaction trace was then analysed by the least squares method.

[B] (μM)	*Actual value of k[B] (sec^{-1})*	*Observed value of k[B] (sec^{-1})*	*Percentage error*
1	20	10.3	48
2	40	30.0	26
3	60	49.7	17
4	80	69.6	13
5	100	89.6	11
10	200	190.3	5
50	1000	995.2	1
100	2000	1977.1	1

A plot of 1/[A] against time will have a slope equal to the second-order rate constant, k (units, M^{-1} sec^{-1}). In this case, the value of $t_{1/2}$ depends on the magnitude of $[A_o]$ and does not remain constant throughout the course of the reaction. This method requires a knowledge of the concentrations of both reactants. The half-times of such reactions are $t_{1/2} = 1/k[A_o]$ and the successive half-times double in length.

5.2.2. *Pseudo-first Order Reactions*

A simpler method is to carry out the second-order reaction under first-order conditions. If the concentration of one of the reactants greatly exceeds that of the other reactant, then its concentration will remain effectively constant during the whole time course of the reaction. For example, if $[A_o] \gg [B_o]$ then

$$\frac{d[AB]}{dt} = k\,[A][B] = k_{obs}\,[B] \qquad \text{Equation 9}$$

where $k_{obs} = k[A_o]$ and so the rate of disappearance of B or formation of AB follows first-order kinetics. The observed first-order rate constant, k_{obs}, can be determined as described in Section 5.1, and the second-order rate constant, k, calculated from $k_{obs} = k[A]$. Using this method, only the concentration of [A] needs to be known accurately.

The question arises as to what '[A] is much greater than [B]' means in practical terms. Often, the concentration of the excess reactant is limited by factors such as the small quantities of material available, the high absorption of the solution, high background of fluorescence or high viscosity of the solution. The effect of the concentration of the excess reagent on the measured rate constant is shown in *Table 3*. It can be seen that ideally at least a 10-fold excess of reactant is required to obtain a reasonably accurate value of the rate constant but a rough value (within 20%) is given by only a 3-fold excess of reactant. However, as the concentrations approach equality, the reaction tends towards second-order which presents another possible error, discussed in Section 6.

5.3 Reversible Second-order Reactions

Many protein–ligand or protein–protein interactions are reversible second-order reactions and so the rate constant for the reverse reaction needs to be considered.

$$A + B \underset{k_{-1}}{\overset{k_1}{\rightleftharpoons}} AB$$

where k_1 is the second-order association rate constant and k_{-1} is the first-order dissociation rate constant.

Again, to simplify the analysis of such a reaction, it is usually carried out with the concentration of one reagent in large excess over that of the other, i.e. $[A_0] \gg [B_0]$. If [B] is the concentration of B at time t; $[B_\infty]$ is the concentration of B at equilibrium; and $[B_0]$ is the concentration of B at time zero; and similar notations are used for the concentrations of A and AB, then at equilibrium

$$k_1 [A][B_\infty] = k_{-1} [AB_\infty]$$

Therefore

$$k_1 [A][B_\infty] - k_{-1} [AB_\infty] = 0 \qquad \text{Equation 10}$$

Since [A] remains constant throughout the reaction, $k_1[A]$ is a constant and is defined here as k'.

Therefore

$$k'[B_\infty] - k_{-1} [AB_\infty] = 0$$

since

$$[B_\infty] = [B_0] - [AB_\infty]$$

$$k'([B_0] - [AB_\infty]) - k_{-1} [AB_\infty] = 0$$

$$k'[B_0] - k'[AB_\infty] - k_{-1} [AB_\infty] = 0$$

$$[B_0] = \frac{k' [AB_\infty] + k_{-1} [AB_\infty]}{k'}$$

$$= \frac{k_{-1}}{k'} [AB_\infty] + [AB_\infty] \qquad \text{Equation 11}$$

The rate of formation of AB is given by the equation:

$$\frac{d[AB]}{dt} = k'[B] - k_{-1}[AB] \qquad \text{Equation 12}$$

Since $[B] = [B_0] - [AB]$

$$\frac{d[AB]}{dt} = k'([B_0] - [AB]) - k_{-1}[AB]$$

$$= k'((k_{-1}/k')[AB_\infty] + [AB_\infty] - [AB]) - k_{-1} [AB]$$

$$= k_{-1}[AB_\infty] + k'[AB_\infty] - k'[AB] - k_{-1} [AB]$$

$$= (k_{-1} + k')([AB_\infty] - [AB]) \qquad \text{Equation 13}$$

This has the same form as a first-order reaction shown in Equation 1. A plot of $\ln([AB_\infty] - [AB])$ against time will therefore yield an observed first-order rate constant, k_{obs}, which is equal to $k_{-1} + k'$. Since $[AB_{\infty}] = [B_o] - [B_{\infty}]$ and $[AB] = [B_o] - [B]$, a plot of $\ln([B] - [B_{\infty}])$ will yield the same information. Again, concentrations need not be in any specific units and any parameter proportional to concentration of the species can be used.

If the reaction is performed over a range of concentrations of A, the dependence of k_{obs} on [A] can be measured. Since

$$k_{obs} = k_1[A] + k_{-1}$$

a plot of k_{obs} against time will give a straight line of slope k_1 (second-order rate constant; units, $M^{-1}\ sec^{-1}$) and intercept on the ordinate of k_{-1} (first-order rate constant; units, sec^{-1}). Hence k_1 and k_{-1} can be determined together with the value of the equilibrium dissociation constant (k_{-1}/k_1) for the reaction.

5.4 Two-step Binding Processes

The rate constant for a diffusion-limited protein–ligand or protein–protein interaction can be measured as described above and is typically in the region of $10^7 - 10^8\ M^{-1}\ sec^{-1}$. This agrees well with theoretical values calculated from a knowledge of diffusion coefficients, dimension of the molecules and viscosity. However, many second-order rate constants for biochemical processes are significantly slower than these values. The simplest interpretation of this is that the interaction is a two-step process consisting of a diffusion-limited second-order step and a first-order isomerisation step.

Although slow second-order rate constants may suggest the presence of a two-step process, more positive evidence can be obtained if it can be shown that the rate of formation of product is not proportional to the concentration of either reactant over a wide range of concentrations. The analysis of such two-step mechanisms becomes complex since the first-order isomerisation step could involve either one of the reactants or the product. Furthermore, it is not usually apparent which step(s) contribute to the observed change of signal. For a detailed example of one approach to this type of mechanism, the reader is referred to Bagshaw *et al.* (15), but one mechanism is discussed here.

Assume that the first-order isomerisation involves the complex of AB and that the initial diffusion-limited reaction is fast compared with the isomerisation

$$A + B \overset{K_1}{\rightleftharpoons} AB \underset{k_{-2}}{\overset{k_2}{\rightleftharpoons}} AB^*$$

The observed rate of formation of AB* is given by

$$k_{obs} = \frac{k_2}{1 + 1/K_1.[B]_o} + k_{-2} \qquad \text{Equation 14}$$

so a plot of k_{obs} against $[B_o]$ gives a hyperbolic relationship (14). Rearranging gives

$$\frac{1}{k_{obs} - k_{-2}} = \frac{1}{k_2} + \frac{1}{K_1k_2\,[B_o]} \qquad \text{Equation 15}$$

A plot of $1/(k_{obs} - k_{-2})$ against $1/[B_o]$ should give a straight line of slope $1/K_1k_2$ and ordinate intercept of $1/k_2$. k_{-2} is determined independently by a displacement reaction (see Section 5.5).

It should be emphasised, however, that this is only one mechanism for a two-step binding process and that the conditions for this analysis may not be possible to achieve in practice. Also it assumes that only AB* is observed, whereas AB may also contribute to the signal.

5.5 Displacement Reactions

The dissociation rate constant of a protein–ligand or protein–protein interaction can often be determined by a displacement reaction. For example

$$AB + C \rightarrow AC + B$$

If a signal change occurs on dissociation of AB or formation of AC, an observed rate constant can be measured when AB is mixed with C.

Two extreme conditions can be considered for this process (16).

$$AB \underset{k_1}{\overset{k_{-1}}{\rightleftharpoons}} A + B$$

$$A + C \xrightarrow{k_2} AC$$

First, B dissociates from A relatively slowly compared with the association rate of C to A. Then the observed rate of formation of AC is constant with respect to the concentration of C if this is high enough to compete effectively with B for unbound A. The observed rate of formation of AC is then limited only by the dissociation rate of AB (i.e. k_{-1}). Second, A and B are in rapid equilibrium and the observed rate of formation of AC increases with the concentration of C. It is important, therefore, to perform experiments over a wide range of concentrations of C in order to determine which situation holds for the reaction being studied. If k_{obs} is independent of [C], then this is the value of k_{-1}. If it is dependent on [C], then it may only be possible to obtain the equilibrium constant for the reaction.

5.6 Competitive Binding Reactions

Even if the binding of a ligand to a protein does not give an observable spectroscopic signal, it may be possible to measure the rate constant for this reaction by performing a competitive binding experiment with a ligand which does give a signal (17).

For example, in the processes:

$$A + B \xrightarrow{k_1} AB$$

$$A + C \xrightarrow{k_2} AC$$

where the association step characterised by k_1 is required but only when the formation of AC gives a signal, the rate of loss of A is given by

$$\frac{-d[A]}{dt} = (k_1[B] + k_2[C])\,[A]$$

If both B and C are in excess over A an equation of the same form as a first-order equation is obtained and so a plot of ln[A] against time gives an observed first-order rate constant of $k_1[B] + k_2[C]$. Since the rate of loss of A is proportional to the rate of formation of AC, a plot of ln[AC] versus time also gives the same information.

The experiment is designed so that one syringe contains A, and the other an excess of B. Then the experiment is repeated with increasing concentrations of C in the syringe containing B. If k_{obs} is plotted against [C], a straight line of slope k_2 and ordinate k_1[B] is obtained.

6. ANALYSIS OF DATA

For the relatively simple reactions discussed above it can be seen that reaction conditions can usually be used that give rise to an exponential process. The first-order rate constant associated with this can be calculated either manually or by computer.

6.1 Manual Methods

A trace of a reaction record can be made on graph paper and the appropriate values of signal against time measured. The suitable logarithmic plot is then made and, from the slope of this, the value of k is determined. This method is somewhat subjective in both the drawing of a smooth curve through a noisy trace and also the determination of the end point of the reaction. As shown below (Section 6.2), the latter parameter is crucial for obtaining an accurate value of k.

6.2 Computer Methods

It is not usually practical to process more than 10 points for the manual determination of rate constants. However, the ready availability of microcomputers makes the processing of several hundred data points relatively straightforward.

Initially, the end point of the reaction is determined by averaging points between certain time intervals selected by the operator. It is extremely important to choose the correct values. Second-order reactions are particularly liable to give errors since the last 10% of reaction proceeds very slowly. The data are then treated as an exponential process so that they are in the form of a linear equation, $y = at + b$. By the method of least squares deviations, the best fit to the data is obtained when $(y_{obs} - y_{calc})^2$ is a minimum and values of a and b are obtained. These are then used to generate a theoretical reaction curve which should be superimposed on the original data. The accuracy of this fit can be judged by eye. However, a better approach is to plot the 'residuals', the differences between the calculated and actual values of y at any time point. They allow one to see readily whether any deviations are the result of random noise or whether they are non-random, indicating that the theoretical equation does not correctly describe the experimental data (*Figure 9*). An introduction to data analysis is given by Edsall and Gutfreund (18). Although they are concerned with the analysis of thermodynamic data, it is equally applicable to kinetic data.

6.3 Simulation of Reactions

Even for the relatively simple reactions described above, it can be seen that the rate equations often cannot be solved analytically and that simplifying assumptions need to be made. Having extracted rate constants it can be instructive to simulate the reaction by computer methods to determine how well the data fits the predicted mechanism. This can be done at two levels; either simulation of a reaction trace or the determination

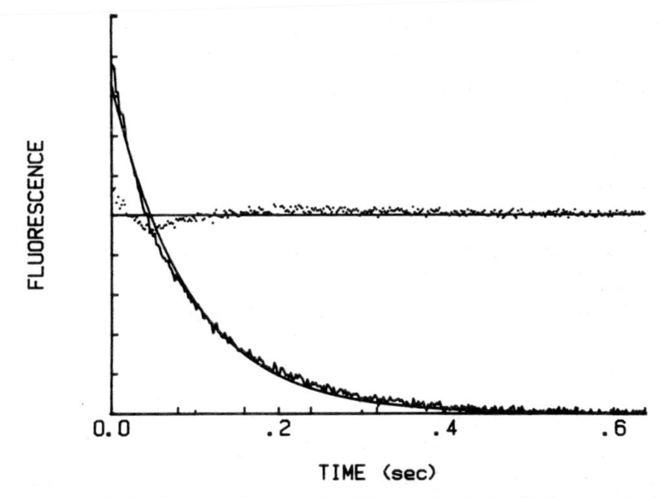

Figure 9. Computer analysis of stopped-flow reaction. The reaction is the displacement by NAD^+ of NADH from alcohol dehydrogenase (ADH) of *Drosophila*. One syringe contained 1.0 μM ADH, 5.3 μM NADH, and the other syringe contained 2.4 mM NAD^+, 5 mM pyrazole (reaction chamber concentrations). Both syringes also contained 50 mM MES, pH 6.5. NADH fluorescence was excited at 340 nm and observed through a Wratten 47B filter. The experimental data points were analysed as a first-order reaction as described in Section 6.2 and the theoretical curve drawn using the values of the constants obtained from this procedure. The central (dotted) line is a plot of the residuals; the differences between the actual and theoretical values. These show that the observed reaction is not first-order under these conditions since the values systematically deviate from zero. (A.R.Place and J.F.Eccleston, unpublished experiment.)

of how an observed process is dependent on the concentration of one of the reactants. Computer simulations are also helpful in the design of experiments and often give a greater insight into a system than can be achieved by intuitive arguments.

The earliest simulations of biological systems were done using analogue computers. Differential equations are solved by the use of voltages as analogues of the variables in the equations. Operational amplifiers, resistors and capacitors are combined to give an output voltage which is proportional to the time integral of the input voltage. Analogue computers have now been superceded by digital computers, although they still have some advantages such as speed.

For details of programs of simulation of biological processes by digital integration methods the reader is referred to a specialised text such as that of Randall (19). The basic principle of the method is that for small increments of time, the rates of change of reactants are assumed to be constant. For example, in the reaction:

$$\mathrm{A} \underset{k_{-1}}{\overset{k_1}{\rightleftharpoons}} \mathrm{B} \xrightarrow{k_2} \mathrm{C}$$

$$\mathrm{dA}/\mathrm{d}t = -k_1[\mathrm{A}] + k_{-1}[\mathrm{B}]$$

$$\mathrm{dB}/\mathrm{d}t = k_1[\mathrm{A}] - k_{-1}[\mathrm{B}] - k_2[\mathrm{B}]$$

$$\mathrm{dC}/\mathrm{d}t = k_2[\mathrm{B}]$$

The computer then calculates the rate of change of each reactant for a large number of small time intervals, and then is usually programmed to plot out a time course of each reactant and intermediate. The operator may have to make a decision as to the

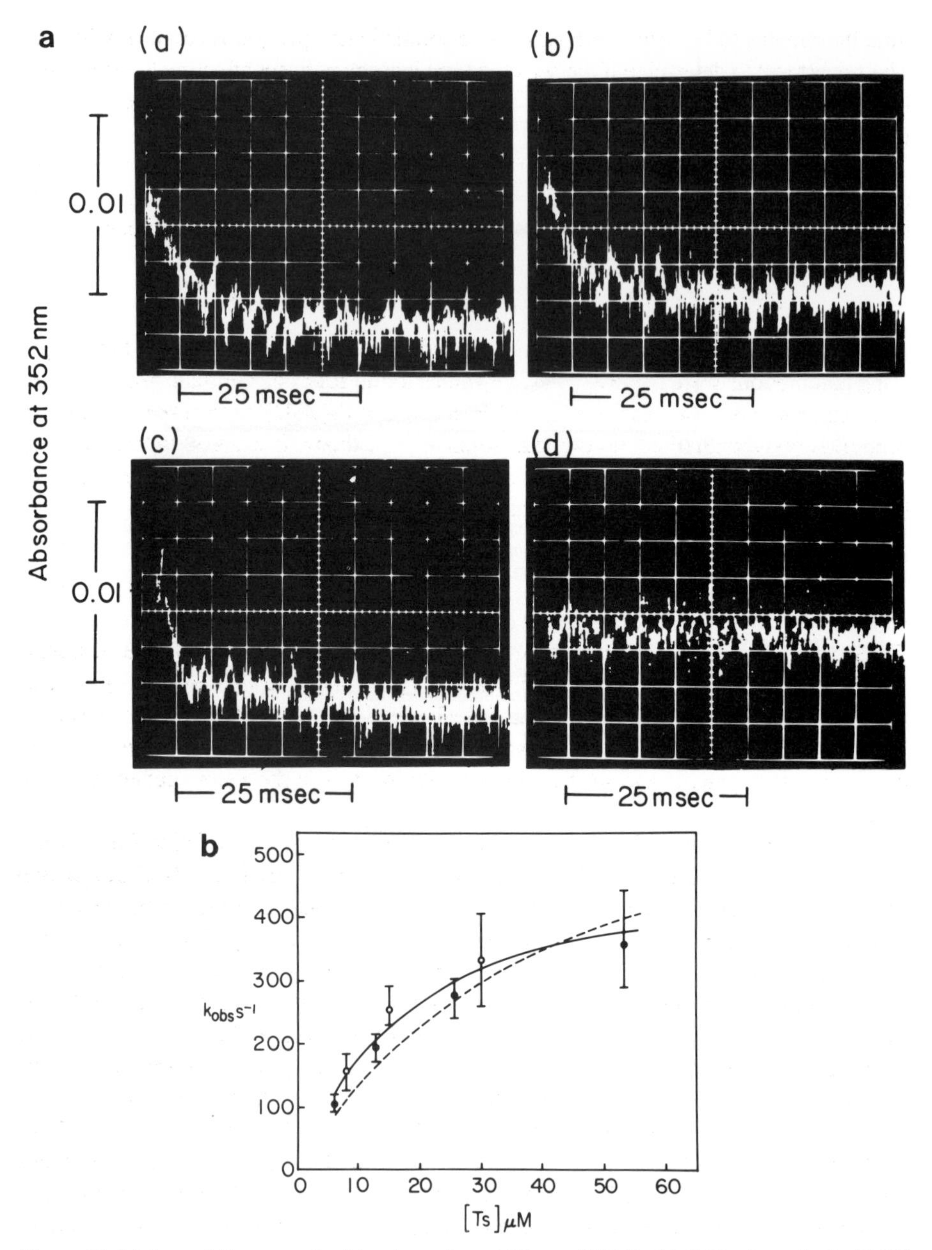

Figure 10. **(a)** Stopped-flow records of the absorption at 352 nm of EF-Tu.thioGDP when mixed with EF-Ts. One syringe contained 3 μM EF-Tu.thioGDP and the other EF-Ts at **(a)** 12.8 μM; **(b)** 53 μM; **(c)** 105 μM; **(d)** 0 μM (reaction chamber concentrations). Both syringes also contained 50 mM Tris-HCl, pH 7.6, 10 mM $MgCl_2$, 0.5 mM dithiothreitol. **(b)** The effect of EF-Ts concentration on the rate constant of the observed process on mixing EF-Tu.thioGDP with EF-Ts. The data points are from the stopped-flow traces in which k_{obs} equals the rate of absorption change analysed as an exponential process. ● and ○ represent data from different preparations of both elongation factors. The dashed line is a computer simulation of the reaction scheme in which $k_1 = 2 \times 10^8$ M^{-1} sec^{-1}, $k_{-1} = 6700$ sec^{-1}, $k_2 = 704$ sec^{-1}, $k_{-2} = 2.2 \times 10^5$ M^{-1} sec^{-1}. The solid line is drawn on the same basis, but $k_1 = 2 \times 10^8$ M^{-1} sec^{-1}, $k_{-1} = 2600$ sec^{-1}, $k_2 = 500$ sec^{-1}, $k_{-2} = 4 \times 10^5$ M^{-1} sec^{-1}. (Ref. 20.)

time increments to be made, although more sophisticated programs can deal with this. The easiest test to determine if the selected time increment is not affecting the simulation is to reduce it by at least a factor of five and to ensure that the same time course of reaction is given.

An example of the use of computer simulations is given in *Figure 10*. The reaction being studied is the displacement of an analogue of GDP from elongation factor Tu by elongation factor Ts (20). This is thought to proceed by the mechanism:

$$\text{Tu.thioGDP} + \text{Ts} \rightleftharpoons \text{Tu.thioGDP.Ts} \rightleftharpoons \text{Tu.Ts} + \text{thioGDP}$$

The rate equations describing this process cannot be solved analytically and so simplifying assumptions were made to calculate values for all four rate constants from stopped-flow experiments. The observed rate of formation of thioGDP with respect to Ts concentration was then studied by computer simulation methods. It can be seen that, although the calculated rate constants give a reasonable fit to the data, the fit can be improved by slight modifications to these values.

7. FUTURE PROSPECTS

The stopped-flow technique was developed in order to observe directly the formation and decay of enzyme–substrate intermediates. Since then the technique has played an important role in elucidating the kinetic mechanism of a wide range of enzymes (13). In recent years, however, attempts have been made to use the technique for investigating more complex systems and in the author's opinion this is the main challenge for the future.

For systems more complex than ligand or substrate binding to small globular proteins, several problems arise. First, it is necessary to be able to prepare sufficient amounts of well-characterised material and to have a suitable spectroscopic probe which can be detected against a possible high background of light scattering. Second, the viscosity of the solution may become too high for rapid mixing to occur. For example, a 5 μM solution of a protein of molecular weight 5×10^4 has a concentration of 0.25 mg/ml, whereas a 5 μM solution of ribosomes (molecular weight 2.5×10^6) has a concentration of 12.5 mg/ml. Third, the complex system may not be stable in the conditions of the stopped-flow experiment. It is known that DNA can be broken into fragments by shear forces produced by rapid flow especially through narrow tubing.

Despite these problems, successful stopped-flow studies have been made on several complex systems. Two are briefly described here. Goss *et al.* (21) have investigated the formation of the *Escherichia coli* protein biosynthesis initiation complex. This involves the binding of the 30S and 50S ribosomal subunits to form a 70S ribosome and the process is mediated by initiation factor 3 (IF-3). The signals that they used were the increase in the light scattering of the solution on the formation of the 70S ribosome and the increase in the fluorescence anisotropy of a fluorescent derivative of IF-3 when it binds to the ribosome. Karpen *et al.* (22) have measured the acetylcholine receptor-mediated ion flux of membrane vesicles. The vesicles were prepared containing the dye, anthracene-1-5-disulphonic acid. The fluorescence of this compound is quenched by Cs^+ ions. Rapid mixing of the vesicles with Cs^+ therefore allowed the ion flux of the ions into the vesicles to be measured.

These two studies well illustrate the potential of stopped-flow techniques for investigating the kinetics of the interactions of complex systems.

8. ACKNOWLEDGEMENTS

I am extremely grateful to Professor H.Gutfreund of the Molecular Enzymology Laboratory at Bristol University for his help and advice over many years on the study of the dynamics of biological processes and on the methods and techniques for their analysis. I also thank Mrs A.Humphrey-Gaskin and Mrs T.Kanagasabai for help in preparing the manuscript.

9. REFERENCES

1. Gutfreund,H. (1972) *Enzymes: Physical Principles.* Wiley-Interscience, London.
2. Hiromi,K. (1979) *Kinetics of Fast Enzyme Reactions.* John Wiley, New York.
3. Dobrowolski,J.A., Marsh,G.E., Charbonneau,D.G., Eng,J. and Josephy,P.D. (1977) *Appl. Optics,* **16**, 1491.
4. Eastman-Kodak Co. (1976) *Kodak Filters for Scientific and Technical Uses.*
5. Haugland,R.P. (1985) *Handbook of Fluorescent Probes and Research Chemicals.* Molecular Probes, Junction City, Oregon.
6. Trentham,D.R., Eccleston,J.F. and Bagshaw,C.R. (1976) *Q. Rev. Biophys.*, **9**, 217.
7. Criddle,A.H., Geeves,M.A. and Jeffries,T. (1985) *Biochem. J.*, **232**, 343.
8. Holbrook,J.J. (1972) *Biochem. J.*, **128**, 921.
9. Rosenfeld,S.S. and Taylor,E.W. (1984) *J. Biol. Chem.*, **259**, 11920.
10. Trentham,D.R., Bardsley,R.G., Eccleston,J.F. and Weeds,A.G. (1972) *Biochem. J.*, **126**, 635.
11. Finlayson,B. and Taylor,E.W. (1969) *Biochemistry,* **8**, 8002.
12. Tsien,R.Y., Pozzan,T. and Rink,T.J. (1982) *J. Cell Biol.*, **94**, 325.
13. Bayley,P., Ahlstrom,P., Martin,S.R. and Forsen,S. (1984) *Biochem. Biophys. Res. Commun.*, **120**, 185.
14. Gutfreund,H. (1975) *Prog. Biophys. Mol. Biol.*, **29**, 161.
15. Bagshaw,C.R., Eccleston,J.F., Eckstein,F., Goody,R.S., Gutfreund,H. and Trentham,D.R. (1974) *Biochem. J.*, **141**, 351.
16. Bagshaw,C.R. and Trentham,D.R. (1974) *Biochem. J.*, **141**, 331.
17. Eccleston,J.F. and Trentham,D.R. (1979) *Biochemistry,* **13**, 2896.
18. Edsall,J.T. and Gutfreund,H. (1983) *Biothermodynamics.* John Wiley and Sons, p. 228.
19. Randall,J.E. (1980) *Microcomputers and Physiological Simulation.* Addison-Wesley Publishing Co., Reading, MA.
20. Eccleston,J.F. (1984) *J. Biol. Chem.*, **259**, 12997.
21. Goss,D.J., Parkhurst,L.P. and Wahba,A.J. (1982) *J. Biol. Chem.*, **257**, 10119.
22. Karpen,J.W., Sachs,A.B., Cash,D.J., Pasquale,E.B. and Hess,G.P. (1983) *Anal. Biochem.*, **135**, 83.

CHAPTER 7

The Determination of Photochemical Action Spectra

DAVID LLOYD and ROBERT I.SCOTT

1. INTRODUCTION

Early observations on the photodissociable nature of carboxyhaemoglobin (1,2) led Warburg (3) to use photodissociation as a means of studying CO-inhibited cellular respiratory systems. The marked inhibition of respiration by CO observed in the dark was relieved on illumination, and an action spectrum was obtained by measuring the effects of different wavelengths at accurately determined light intensities. This technique enabled the unequivocal demonstration that the Atmungsferment or 'oxygen-transferring enzyme' was a haemoprotein with absorption maxima at 590 and 433 nm (4,5) and that similar chromophores were present in both yeast and heart muscle (6–8). Different results with *Acetobacter pasteurianum* (5,9) first showed that a bacterial terminal oxidase was different from those of yeast and muscle. Studies of the effect of CO on aerobic ethanol fermentation in yeast (6) and in rat retina (10), gave action spectra for the 'Pasteur enzyme'. These spectra were similar, although not identical with those for respiration. On the basis of these action spectra it is assumed that the pigment responsible for the Pasteur Effect and cytochrome oxidase are not distinct (11,12).

The measurement of photochemical action spectra has proved to be of key importance in many fields of biochemical research. Such spectra identify the CO-combining microsomal pigment cytochrome P-450 as the terminal oxidase of mixed-function oxidase systems (13–15), phytochrome, the photosensitive pigment controlling development in higher plants (16,17), and the photosynthetic pigments of algae (18,19).

2. APPARATUS

The basic apparatus required for determining the photochemical action spectrum of a CO-inhibited respiratory system comprises a chamber in which the rate of oxygen consumption can be recorded in the presence of CO and in the presence and absence of light of an appropriate wavelength (see *Figure 1*).

Many of the early attempts to apply the method to bacteria met with limited success (12,20,21). Even where marked inhibition of respiration by CO was observed, it was not always possible to demonstrate reversibility in strong light even at low temperatures (light sensitivity increases with decreasing temperature). Technical problems inherent in determining photochemical action spectra were considerably eased by use of an oxygen electrode instead of the classical manometric method (22–24) for measurement of respiration. The improved methodology enabled Castor and Chance (25) to demonstrate

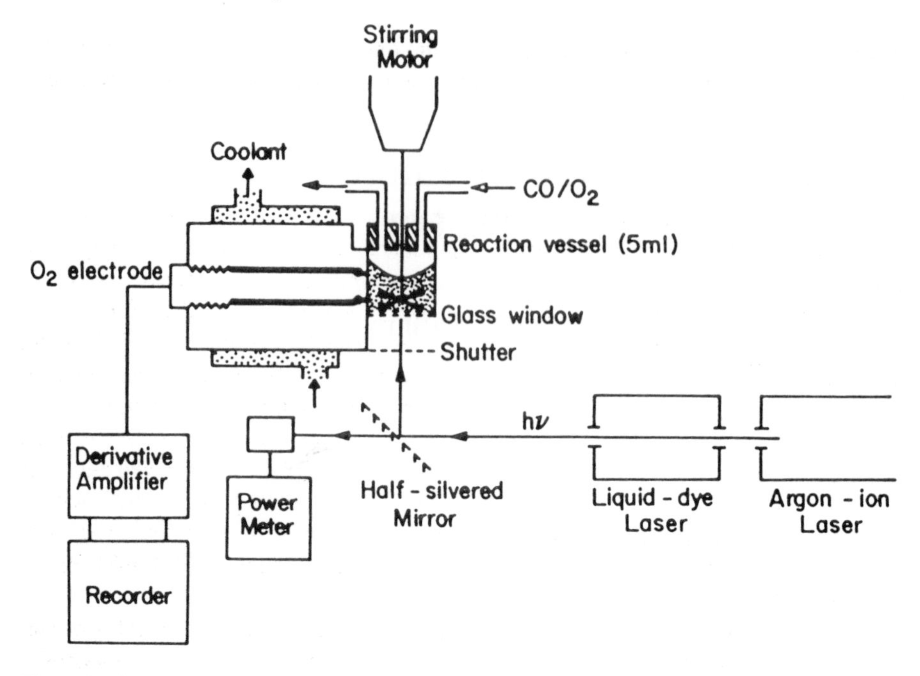

Figure 1. Illumination of the reaction vessel by light from the liquid dye laser.

the functional roles of cytochrome *o* and a_2 (now called cytochrome *d*) in different bacterial respiratory systems.

Despite the success of the method developed by Castor and Chance (25) the technique has the following practical problems.

(i) Continuous drift of electrode current.
(ii) Slow response makes it difficult to obtain a complete spectrum from a single drop.
(iii) Evaporation of the suspension changes the geometry of the gas–liquid interface: gas exchange rates alter during measurement of the spectrum.
(iv) Settling of cells within the drop.
(v) Gradients of gases in the unstirred drop.
(vi) Effect of light on the electrode.

Hyde (26) replaced the hanging drop by an annular drop, using an electrode cap to shield it from light, and reduced the time required to obtain spectra by taking the first derivative of the oxygen-electrode response. This instrument used a Zeiss Xenon source and a Bausch and Lomb monochromator of 6.4 nm/mm dispersion and 500 nm blaze. Slits were adjusted to provide incident light with spectral band width (see Chapter 2) at half peak height of 2 or 4 nm.

2.1 Current Instrumentation

A block diagram of a modern apparatus for recording photochemical action spectra is shown in *Figure 1*. The first published spectra obtained with this version of the apparatus were those of Edwards *et al.* (27) and Edwards and Chance (28) with bacterial *a* and *o*-type oxidases and the cytochromes a_3 and *o* of the trypanosomatid *Crithidia oncopelti* which harbours a bacterial endosymbiont. The recent demonstration of a cytochrome *o* as terminal oxidase (and receptor for aerotaxis) in *Salmonella typhimurium* (29) employed the same equipment.

2.1.1 *Light Source*

A tunable dye laser (D.Lloyd, S.W.Edwards and B.Chance, unpublished results) provides the opportunity for investigation of terminal oxidases, which in the CO-liganded state are refractory to photodissociation, even at the highest light intensities available with conventional Xenon sources coupled with a monochromator. The narrow line-width (<0.01 nm) and high intensity of the laser light enables extremely detailed spectra to be obtained. Using conventional monochromators action spectra have been obtained at bandwidths of 2 to 4 nm for *Crithidia fasciculata* (30) but the high intensities needed for photolysis of the CO-liganded oxidases of *Azotobacter vinelandii* necessitated the use of much wider slit-widths (e.g. 13 nm) (24). Hoffman *et al.* (31) showed that laser power giving half maximum light relief of CO-inhibited respiration mediated by cytochrome *d* in *A. vinelandii* was 68 mW/cm^2 at 637 nm; this was much higher than for cytochrome *a* (13.5 mW/cm^2 at 590 nm) or for cytochrome *o* (3.8 mW/cm^2 at 560 nm).

The output of an 8-W argon-ion laser (Lexel, Model 95) drives the liquid dye laser (Spectra Physics, Model 375). Sufficient power (up to 2.0 W) for excitation of rhodamine 6G (572–632 nm; output up to 200 mW) is available using the 514.5 nm line of the argon-ion laser. Multiline operation (using a Model 502 multiline mirror holder A) gives more than 3.0 W output which is necessary to excite rhodamine 110 (532–572 nm). Wavelengths longer than 632 nm may be obtained using rhodamine B, *Table 1* lists suitable dyes and their working wavelength range. A half-silvered mirror splits the beam of lased light between the stirred cell suspension and the attenuator of a power meter (Spectra Physics, Model 404). An electrically-operated shutter is placed between the beam-splitting mirror and the glass bottom of the reaction vessel (*Figure 1*). Power output of the dye laser is adjusted after each change of wavelength by altering the power output of the pump (argon-ion) laser so that measurements over the entire tunable range are made at equal incident energies.

2.1.2 *Oxygen Electrode Chamber*

The rapid acquisition of action spectra is greatly facilitated by the use of a stirred cell

Table I. Wavelength ranges of some dyes suitable for liquid dye lasers.

Dye[a]	*Wavelength range (nm)*	*Peak power*[b] *(mW)*
Rhodamine 110	532–572	200
Rhodamine 6G	572–632	800
Rhodamine B	632–660	200
Coumarin 7	510–570	100
Coumarin 120	314–475	75

[a]'Laser' quality dyes and solvents are essential to extract optimum performance from liquid dye lasers. Take great care not to contaminate one dye with traces of other dyes left in the laser pump from previous runs.
[b]Maximum power available using a Lexel Model 95 Argon ion laser to drive a Spectra Physics Model 375 liquid dye laser.

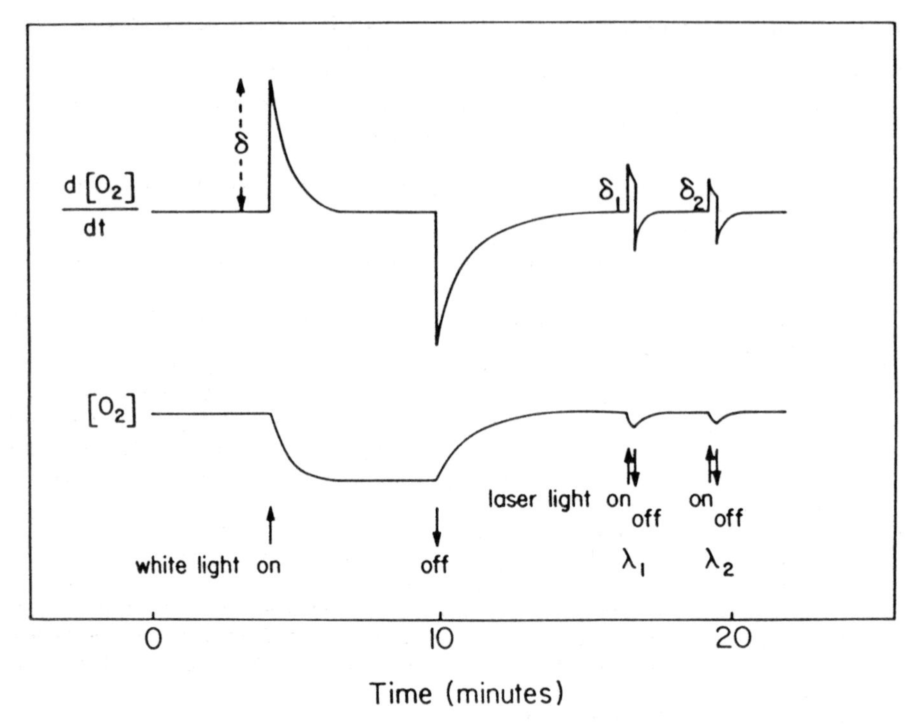

Figure 2. Typical traces obtained for O_2 concentration and its first derivative during alternating phases of darkness and illumination by white light and by the dye laser. See *Table 2* for a detailed experimental protocol.

suspension in a system open to a mobile gas phase (32) (*Figure 1*). This arrangement (33) allows precise control of oxygen and CO concentrations in the cell suspension over extended times and permits the determination of action spectra both at defined O_2 concentrations and over a wide range of accurately defined $CO:O_2$ ratios. Furthermore the stirred vessel allows measurements to be made at a variety of temperatures not easily accessible with unstirred drops and is easily applicable to fragile cell-types such as amoebae or cultured mammalian cells.

The cylindrical open reaction system (5.0 ml working volume) is constructed of stainless steel with a side port (9.5 mm diameter) and a glass bottom. Gas entry and exit ports are via the lid, and the vertically-mounted stirring shaft is driven at up to 2800 r.p.m. by a tacho-controlled motor with digital display of stirring speed. The vessel is jacketed and supplied with water from a constant-temperature bath. Mixing of O_2 and CO employed a digital gas mixer (34) which provides accurate dilution in 5% steps; usually the $CO:O_2$ ratio employed was 19:1, but the facility for alteration of this ratio is a valuable one in those cases where the effect of steady state dissolved O_2 on oxidase function is to be assessed. The oxygen electrode was Type E5047 (Radiometer A/S, Copenhagen, Denmark) fitted with a 12 μm polypropylene membrane. The output was monitored on a three-channel potentiometric recorder after amplification (Johnson Foundation oxygen electrode amplifier fitted with pre-amplification stage and first-order derivative output). Simultaneous recording of the electrode output (dissolved oxygen) and its derivative were always employed; the third recorder channel was for laser energy. Use of a derivative amplifier enables rapid and precise measurements of respiration rates.

Table 2. Experimental procedure to establish conditions suitable for obtaining an action spectrum.

Figure 2 illustrates typical experimental records of the following protocol.

1. Equilibrate the reaction vessel (see *Figure 1*) at a preselected temperature; select the composition of the mobile gas phase (typically N_2:O_2, 19:1); add the experimental medium (5 ml) to the reaction vessel and switch on the stirrer.
2. After temperature equilibration, switch on the oxygen electrode circuit and select amplification settings appropriate to the potentiometric recorder.
3. Add the organisms/preparation: the oxygen electrode trace (labelled [O_2] in *Figure 2*) will settle to a new steady state level due to the consumption of some of the oxygen supplied. The shutter should be closed.
4. Substitute carbon monoxide for nitrogen at the input of the gas mixer: the steady state concentration of oxygen will increase due to the inhibition of respiratory oxygen consumption.
5. Switch on the derivative amplifier (this provides the trace labelled d[O_2]/dt in *Figure 2*).
6. Open the shutter to a beam of white light provided, for example, by a 150 W quartz-iodide tungsten lamp placed 10 cm from the sample ('white light on' in *Figure 2*): the steady state level of dissolved oxygen (lower trace) moves to a new (lower) level as inhibition by CO is relieved. The new rate of respiration is indicated by the initial excursion (δ) of the derivative (upper) trace and, eventually, by the position of the lower, [O_2] trace.
7. Close the shutter ('white light off'): the system returns to the former 'dark' steady state.
8. Adjust the amplifier settings to give adequate changes in amplitude on the recorder.
9. Tune the dye laser output to the desired wavelength (λ_1).
10. Select and record a light intensity below the saturation value. Open the shutter ('laser light on'): the initial excursion of the derivative trace gives the extent of the relief of respiratory inhibition. (δ_1 arbitrary units). Close the shutter as soon as the derivative trace indicates a decreasing rate of change ('laser light off').
11. Allow the 'dark' steady state to re-establish.
12. Tune the laser to a second (λ_2) and, subsequently, further wavelengths within the wavelength range of the dye laser and repeat step 10. Remember to adjust the dye laser output to constant incident intensity at each wavelength.
13. The action spectrum is a plot of δ_n versus λ_n.

2.1.3 *A Worked Example*

Figure 2 illustrates typical traces obtained for oxygen concentration and its first derivative during alternating phases of darkness and illumination by the liquid dye laser using the protocol itemised in *Table 2*.

In this experiment the gas mixture was adjusted to 19:1, CO:O_2. Buffer (4.5 ml of 50 mM K-phosphate at pH 6.8) was equilibrated at 25°C under this gas whilst stirring at 1500 r.p.m. A dense suspension of organisms was added.

Steady state dissolved oxygen at the required level is first established in the dark: this can be adjusted by altering: (i) cell concentration, (ii) composition of the mobile gas phase, or (iii) the stirring rate. Exposure to laser light of appropriate wavelength leads to rapid photolytic decomposition of the CO-liganded oxidase and the consequent increased respiration resulted in a monotonic transition to a lower steady state level of oxygen dissolved in the cell suspension. The attainment of the new steady state is a slow process (typical $t_{1/2}$ values 2–3 min) limited by oxygen diffusion across the interface of the stirred suspension. The initial velocity of approach to the new steady state is an accurate measure of the eventual displacement of oxygen concentration, and is instantaneously recorded by the maximum excursion of the derivative trace. The action spectrum is then the plot of this maximum excursion (labelled δ in *Figure 2*) versus

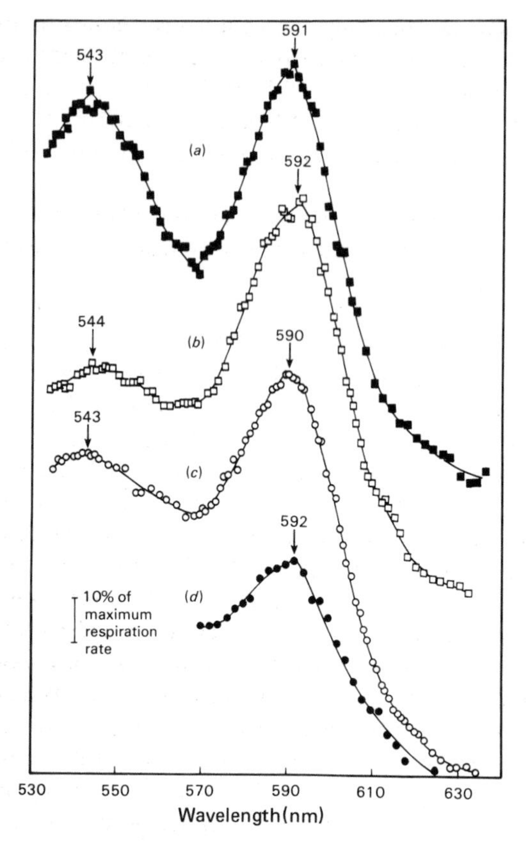

Figure 3. Photochemical action spectra of intact cells of *A. castellanii*. Cells were harvested during (**a**) the early-exponential, (**b**) the late-exponential or (**c**) the stationary phases of growth. (**d**) Shows the action spectrum of the late-exponential phase cells in the presence of 1 mM azide. Laser powers used to illuminate the cell suspensions for Rhodamine 6G were 30 mW for each of (**a**), (**b**), (**c**) and (**d**); for Rhodamine 110 (532–572 nm) were normalised to those of Rhodamine 6G (572–633 nm) by assuming the rates obtained at 572 nm with the two dyes to be equal; respiration rates were then expressed as a function of the maximum rate obtained. The O_2 concentration in the liquid phase was (**a**) 1.5 μM, (**b**) 3.7 μM, (**c**) 3.8 μM and (**d**) 4.5 μM. The gas phase passing over the cell suspension was CO/O_2 (19:1), which is equivalent to concentrations of 814 μM-CO and 56 μM-O_2. Reproduced with permission from reference 35.

wavelength, normalised to constant incident energy (see *Figure 3*). The wavelength scale is calibrated with a Hartridge reversion spectroscope checked against a sodium lamp (Na D emission line at 589.3 nm). Restoration to the dark steady state occurs on closing the shutter.

2.1.4 *Artefacts*

Reversal of respiratory inhibition by CO is dependent upon the power and wavelength of the light used for irradiation. In preliminary experiments establish that the respiratory response has a linear dependence on incident intensity; within this 'working range' light saturation (typically observed above ~35 mW) is avoided. Possible bleaching of absorption bands by lased light did not prove to be a problem; similar responses were recorded before and after 50 determinations in our applications. This should be checked routinely.

Sensitivity of oxygen electrode output to light has sometimes necessitated elaborate precautions for protection of its surface from the beam (26). In the experimental set-up described here the minute cathode area of the Radiometer electrode and its orientation normal to the incident illumination make light-induced current changes insignificant. The heating effect of the illumination is also negligible in terms of its effect on respiratory rates.

In some systems where photolysis of CO-liganded oxidases requires high intensities (25) increased response of inhibited respiratory systems is produced by lowering the $CO:O_2$ ratio or by lowering the temperature.

3. APPLICATIONS AND FUTURE PROSPECTS

Examples of spectra obtained from the soil amoeba, *Acanthamoeba castellanii* (*Figure 3*) illustrate that the only detectable functional haemoprotein oxidase present at any stage of growth in this organism is cytochrome a_3, with action spectra α and β maxima at 590–592 and 543–544 nm, respectively (35). Similar results for *Crithidia fasciculata* (36), *Leishmania tarentolae* (37) and *Saccharomyces uvarum* (33) clearly indicate that the CO-reacting *b*-type haemoproteins in these lower eukaryotes are not functional terminal oxidases. A distinctive terminal oxidase is present in the ciliated protozoon, *Tetrahymena pyriformis*: this is cytochrome a_{620} (38–40) which shows maxima in action spectra at 597 and 547 nm (41). Growth conditions (with added iron or in the presence of an uncoupler of energy conservation) leading to marked decreases in the cyanide-sensitivity of respiration in this organism, gave no marked alteration in the action spectra (42). These data confirm that alternative electron transport chains in all five lower eukaryotes utilise non-cytochrome terminal oxidases of as yet uncertain identities. The simultaneous co-functioning of cytochromes a_3 and *o* in a thermophilic bacterium, PS3 has been confirmed by this method (43): α-absorption maxima were at 588 and 568 nm respectively.

A recent further modification of the photochemical action spectrum apparatus has replaced the oxygen electrode with a membrane-covered mass spectrometer probe (44). This detector enables simultaneous measurement of respiration (O_2 consumption) and glycolysis (CO_2 output) and hence the inactivation of glycolysis by oxygen (Pasteur Effect). Resuts with *S. uvarum* suggest that the action spectrum for the Pasteur Effect is not identical with that of respiration. In addition to the α-maximum of cytochrome a_3 at 590 nm, a second unassigned maximum at about 582 nm was observed (unpublished experiments).

The availability of a wide variety of laser dyes (e.g. the coumarins; see *Table 1*) makes possible the determination of action spectra at other wavelength ranges.

4. ACKNOWELDGEMENTS

The authors wish to thank Dr B.Chance for his continuing interest and encouragement and the World Health Organization, The Royal Society and Science and Engineering Research Council for financial support.

5. REFERENCES

1. Haldane,J. and Smith,J.L. (1896) *J. Physiol.*, **20**, 497.
2. Hartridge,H. and Roughton,F.J.W (1923) *Proc. Roy. Soc. B*, **94**, 336.

3. Warburg,O. (1926) *Biochem. Z.*, **177**, 471.
4. Warburg,O. and Negelein,E. (1928) *Biochem. Z.*, **202**, 202.
5. Warburg,O. and Negelein,E. (1929) *Biochem. Z.*, **214**, 64.
6. Melnick,J.L. (1941) *J. Biol. Chem.*, **141**, 269.
7. Melnick,J.L. (1942) *J. Biol. Chem.*, **146**, 385.
8. Philips,F.S. (1942) *Fed. Proc.*, **1**, 129.
9. Kubowitz,F., Haas,E. (1932) *Biochem. Z.*, **255**, 247.
10. Stern,K.G. and Melnick,J.L. (1941) *J. Biol. Chem.*, **139**, 301.
11. Dickens,F. (1951) In *The Enzyme*, Vol. **2**, Sumner,J.B. and Myrback,K. (eds.), New York, Academic Press, pp. 624.
12. Keilin,D. (1970) *The History of Cell Respiration and Cytochrome*. Cambridge University Press.
13. Estabrook,R.W., Cooper,D.Y. and Rosenthal,O. (1963) *Biochem. Z.*, **338**, 741.
14. Cooper,D.Y., Levin,S., Narasimhulu,S., Rosenthal,O. and Estabrook,R.W. (1965) *Science*, **147**, 400.
15. Rosenthal,O. and Cooper.Y. (1967) in *Methods in Enzymology*. Estabrook,R.W. and Pullman,M.E. (eds), Academic Press, New York, Vol. **10**, p. 616.
16. Parker,M.W., Hendricks,S.B., Borthwick,H.A. and Scully,N.J. (1946) *Bot. Gaz.*, **108**, 1.
17. Seigelman,H.W. and Hendricks,S.B. (1964) *Adv. Enzymol.*, **26**, 1.
18. Haxo,F.T. and Blinks,L.R. (19500) *J. Gen. Physiol.*, **33**, 389.
19. Warburg,O., Krippahl,G. and Schroder,W. (1955) *Z. Naturforsch.*, **106**, 631.
20. Negelein,E. and Gerischer,W. (1934) *Biochem. Z.*, **268**, 1.
21. Chaix,P. and Fromageot,C. (1942) *Trans. Soc. Chim. Biol.*, **24**, 1128.
22. Chance,B., Smith,L. and Castor,L.N. (1953) *Biochim. Biophys. Acta*, **12**, 289.
23. Castor,L.N. (1954) *Determination of Photochemical Action Spectra of Carbon Monoxide-inhibited Respiration*. Ph.D. Thesis, University of Pennsylvania.
24. Castor,L.N. and Chance,B. (1955) *J. Biol. Chem.*, **217**, 453.
25. Castor,L.N. and Chance,B. (1959) *J. Biol. Chem.*, **234**, 1587.
26. Hyde,T.A. (1967) *Design and Use of a New Instrument for the Investigation of Photochemical Dissociation of Carbon Monoxide Inhibited Respiration*. M.Sc. Thesis, University of Pennsylvania.
27. Edwards,C., Beer,S., Sivarum,A. and Chance,B. (1981) *FEBS Lett.*, **128**, 205.
28. Edwards,C. and Chance,B .(1982) *J. Gen. Microbiol.*, **128**, 1409.
29. Laszlo,D.J., Fandrich,B.L., Sivaram,A., Chance,B. and Taylor,B.L. (1984) *J. Bacteriol.*, **159**, 663.
30. Kusel,J.P. and Storey,B.T. (1973) *Biochim. Biophys. Acta*, **314**, 164.
31. Hoffman,P.S., Irwin,R.M., Carreira,L.A., Morgan,T.V., Ensley,B.D. and Der Vartanian,D.V. (1980) *Eur. J. Biochem.*, **105**, 177.
32. Degn,H., Lundsgaard,J.S., Petersen,L.C. and Ormicki,A. (1980) *Methods Biochem. Anal.*, **26**, 47.
33. Lloyd,D. and Scott,R.I. (1983) *Anal. Biochem.*, **128**, 21.
34. Lundsgaard,J.S. and Degn,H. (1973) *IEEE Trans. Bio-Med. Eng.*, **20**, 384.
35. Scott,R.K. and Lloyd,D. (1983a) *Biochem. J.*, **210**, 721.
36. Scott,R.I., Edwards,S.W., Chance,B. and Lloyd,D. (1983) *J. Gen. Microbiol.*, **129**, 1983.
37. Scott,R.K. and Lloyd,D. (1983b) *Soc. Gen. Microbiol. Qt.*, **10**, M19.
38. Turner,G., Lloyd,D. and Chance,B. (1971) *J. Gen. Microbiol.*, **65**, 359.
39. Lloyd,D. and Chance,B. (1972) *Biochem. J.*, **128**, 1171.
40. Kilpatrick,L. and Erecińska,M. (1977) *Biochim. Biophys. Acta*, **460**, 346.
41. Lloyd,D., Scott,R.I., Edwards,S.W., Edwards,C. and Chance,B. (1982) *Biochem. J.*, **206**, 367.
42. Unitt,M.B., Scott,R.I. and Lloyd,D. (1983) *Comp. Biochem. Physiol.*, **74B**, 567.
43. Poole,R.K., Scott,R.I., Baines,B.S., Salmon,I. and Lloyd,D. (1983) *FEBS Lett.*, **150**, 281.
44. Lloyd,D. (1985) In *Gas Enzymology*. Degn,H., Cox,R.P. and Toftlund,H. (eds.), Reidel, pp. 37.

INDEX

Absorbance,
calibration, 29, 30
definition, 6
molar, 7, 16
spectrum, 7
ABTS, 69
Accuracy, 51
Acridine dyes, 116
9-amino acridine, 16, 34, 36, 121–122, 133
acridine orange, 122, 133
and endosome pH, 133
Actomyosin, 111, 112
Adenosine analogues, 95, 141–142, 148
Aequorin, 73
Air bubbles, 9, 19, 149
Albumin,
binding of dansylglycine, 152
in protein calibration, 56
Alcohol dehydrogenase, 161
Allostery, 96
p-Aminobenzoyl glutamate, 105
2-Amino 6-mercaptopurine riboside, 141
Amplification (assays), 55, 73, 77
Anilinonaphthalene sulphonate, 94, 96
Anthracene-1,5-disulphonate, 163
Antibodies, 20, 22, 67, 94
Antifoam, 44
Ascorbic acid, 79, 84, 85,152
Assay,
ADP, 77, 80
cAMP, 77
ATP, 50, 54, 72, 74, 77, 80
ATPase, 83
Biuret, 53, 57
coupled, 81, 86
creatine phosphate, 77
cycling, 55, 77
DNA, 64, 65
discontinuous, 81, 83
end point, 54, 67
enzyme, 80 ff.
fructose, 72
glucose, 69 ff.
luminescent, 73
NAD(P), 67, 72, 73, 78, 141–144
NAD(P)-linked, 67, 70, 71, 88
protein, 53 ff.
RNA, 65
specificity, 50, 51
sucrose, 72
Autofluorescence, 132
Bacteriochlorophyll, 50, 126
Bandwidth, 31
natural, 32, 99
spectral, 32, 166
Baseline correction, 27
Beer–Lambert law, 7, 15, 30, 49
Benzyl orange, 144
Bicinchoninic acid, 60, 61
Binding stoichiometry, 51, 91
Bioluminescence, 73
Bromochlorophenol blue, 144
Bromocresol green, 144
Bromothymol blue, 144

Ca^{2+} ions, 116
and actomyosin, 111–112
and aequorin, 73
and Quin-2, 145–146
and calmodulin, 146
Carbon monoxide, 29, 43, 137, 165
photodissociation, 29, 44, 165–172
Carotenoid, 23, 115, 116, 119, 125–127
Catalase, 42, 70
Cavitation, 139, 149
Cerebral cortex, 123–124
Chemiluminescence, 73, 153
7-Chloro 4-nitrobenz 2-oxa 1,3-diazole, 95
Chlorophenol red, 144
Choroplast, 50, 115
Chlorophyll, 50, 115
Chopper, 3, 4, 5
Chromatophore, 50, 119, 125
Chromogenic substrate, 81
Citrate, 85, 88
Colour development, 51
Continuous flow, 138
Coomassie blue, 61
Coumarin dye (for lasers), 167, 171
Coupled assay, 81, 86
Creatine phosphate, 77
Cresol red, 144
Cuvette, 39
cleaning, 20, 40, 53, 69, 133
flow through, 12
fogging, 9, 15
low temperature, 38, 41
matching, 27, 99
material, 9, 40, 97
pathlength, 9, 99
semi-micro, 9, 39, 53
tandem, 98
Cyanide, 44, 88
Cyanine dye, 22, 118, 121, 128
Cytochrome, 115–116, 166–167

absorbtion bands, 41
bacterial, 28, 29, 166–167
c, 34, 35, 36, 45
d, 43
mitochondrial, 123–124
oxidase, 38, 42, 44, 123–124
P450, 165

Dansyl, 94, 95
excitation filters, 140
Dark current, 8
Dead time, 139, 149–152
Deoxycholate, 58, 60, 62
Deproteinisation, 67
Dewar, 37
2,6-Dichlorophenolindophenol (DCPIP), 152
Didymium, 29
Digitisation, 47, 141
2,4-Dinitrophenol (for dead time determination), 151–152
Dispersion, 32
Dithionite, 41
Double beam, 3, 25, 97, 98
Dry weight, 55, 56
Dual beam, 5
Dual wavelength, 6, 25, 29, 47

Elongation factor, 161–162
Emission spectra, 13, 23, 30, 100
End point,
assay, 54, 67
determination, 152, 160
Enhancement,
absorbance, 35, 37, 86
fluorescence, 64, 65, 92, 96, 102
Enzyme assay,
continuous, 80
coupled, 81, 86
discontinuous, 81
regenerating, 77, 84, 86
Ethidium bromide, 65
Extinction coefficient, 7, 15

FCCP, 122, 129, 131
Ferricyanide, 42
Filter, 2, 13, 18, 140
neutral density, 30
First order reaction, 154
Flavoprotein, 25, 39, 115, 120
Fluorescamine, 63
Fluorescein, 94, 95, 131
Fluorescence,
anisotropy, 20, 94
contamination 18, 40, 100
emission, 13, 20, 33
enhancement, 90, 96, 102
excitation, 13
flavoprotein, 39, 120
front face, 18, 117, 121
lifetime, 2, 20
microscope, 15, 20, 118, 132
polarisation, 20
protein, 95, 105, 106
surface (tissue), 120
quenching, 19, 20, 142
titration, 96, 98
Folin-Ciocalteau reagent, 59
Formycin, 94, 95, 141, 146

Gain, 18, 24, 31
Gel scanner, 12
Glucose oxidase, 69, 70
Glucose-6-phosphate dehydrogenase, 54, 70, 71, 78, 81, 88
Glutamate dehydrogenase, 78
Glyceraldehyde-3-phosphate dehydrogenase, 142–144
Glycerol, 33, 37, 59, 75
Guggenheim plot, 155

Haemoglobin, 44, 91, 116, 119, 123–124, 137, 165
Haemoproteins, 35, 38, 40, 43, 44, 93, 165, 171
see also Cytochromes
Half time, 154–155
Henderson-Hasselbalch equation, 132
Heparin, 65, 66
Hexachloroiridate, 42
Hexokinase, 71
Histones, 64, 65
Hoechst 33258, 64
Holmium, 29
H.p.l.c., 49, 50
Hydrogen peroxide, 42, 69, 73, 85, 86

Indicator reaction, 81
Initiation factor, 163
Inner filter effect, 16, 17, 92, 99, 102, 121, 146
correction, 105
Internal reflection, 37
Internal standard, 65, 66, 75
Intracellular cations, 130
Isosbestic point, 28–29, 149
Ionophore, 120, 125
FCCP, 122, 129, 131

monensin, 132
uncoupler, 125
valinomycin, 122, 125–130

K_d estimation, 107
K_m, 68, 80, 81

Lactate dehydrogenase, 86, 92, 100, 101, 142–144
Lamp
deuterium, 2, 12, 29, 31, 97, 139
mercury arc, 2, 13, 31, 139
temperature, 7
tungsten-halogen, 2, 7, 12, 97, 139
tungsten-iodide, 31
xenon, 2, 13, 31, 139, 167
Laser, 167, 171
Lettre cells, 128, 131–134
Light guide, 12, 117–118, 140
Linearity, 8, 83
Low temperature,
accessory, 37
fluorescence, 39
spectrum, 35, 123–124
Luciferase,
firefly, 50, 54, 74
bacterial, 73
Luminol, 73, 153
Lymphocyte, 129

Malachite green, 86
Membrane potential, 124 ff.
indicators, 115, 119
Lettre cells, 128
lymphocyte, 150
null point method, 128–130
4-Methyl umbelliferone, 150
Mg^{2+} ions, 116
Microscope, 12
fluorescence, 15, 21, 118, 132
Mitochondria, 38, 39, 44, 119, 127
Mixing, 39–40
efficiency, 149–150
in stopped flow, 137–138
gases, 168
Molar absorbtivity, 7, 16
Monensin, 132
Monochromator, 2, 13, 24
calibration, 12, 29
Myoglobin, 119
Myosin, 92, 93, 96, 145, 148

Neutral red, 116, 131–134
p-Nitrophenol, 144
Non-linearity, 8, 16, 52, 83, 84, 106, 109
Null point method, 128–130

Oil, for cation determination, 130
Oxonol dyes, 116, 126, 128
Oxygen, 42, 43, 69, 73, 119
electrode, 168, 171

Pasteur effect, 171
Peptides, 63
Peroxidase, 70
Persulphate, 42
pH,
indicators, 116, 122, 144
of cells, 131–134
of endosomes, 131–132
temperature dependence, 37
Phenolphthalein, 144
Phenol red, 144, 150
6-Phosphogluconate dehydrogenase, 78, 79
Phosphomolybdate, 51, 53, 59, 60, 83, 86
Photobleaching, 1, 19, 97, 119, 170
Photodecomposition, 97, 99
Photodissociation, 38, 44, 165 ff.
Photolysis, 29, 44
Photomultiplier, 3
Phytochrome, 165
Plasticisers, 40
Potassium chromate, 30
Potentiometric titration, 40
Precision, 51
Protein assay,
Biuret, 53, 57
calibration, 56
dye binding, 61
fluorescamine, 63
Lowry, 53, 59
sensitivity, 56
Pulsed light source, 13, 19, 97
N-(1-pyrenyl)iodoacetamide, 142
N-(1-pyrenyl)maleimide, 95
Pyridine nucleotide fluorescence (in tissue), 115, 120
see also Assay, NAD(P)
Pyruvate kinase, 84, 88, 144

Quantum yield, 2, 15
Quene-1 (cytoplasmic pH indicator), 116, 133
Quin-2 (Ca^{2+} indicator), 145–146

Rate constant,
first order, 154, 156
second order, 156, 158

Ratio recording, 15, 30, 97
Raman scattering, 15, 19
Rayleigh scattering 15, 19
Redox titration 40
Reducing sugar, 57, 67
Reflectance,
  specular, 12, 123
  tissue, 123–124
Regenerating enzyme, 86, 88
Resolution, 31, 46
Response time, 33, 36, 41
Rhodamine, 167
Rhodopseudomonas capsulata, 119
Rhodopseudomonas spheroides, 119, 125–126
Rhodospirillum rubrum, 50

Scan speed, 33, 41
Scatchard plot, 108, 110
Scattered light, 101, 120, 140
Scintillation counter, 73, 75
Second order reaction, 145, 147
  rate equations, 155–160
Semicarbazide, 67
Signal/noise ratio, 9, 33, 42, 120
Single beam, 3, 8, 15, 25, 26, 30
Single turnover, 111, 125, 148
Sipper, 12, 14
Slitwidth, 8, 18, 24, 31, 32, 41, 97, 98, 101
Sodium potassium tartrate, 57, 59
Specific activity, 53, 80
Spectrum,
  absolute, 13, 25
  absorbtion, 7
  action, 165 ff.
  CO difference, 43
  corrected, 13, 30
  derivative, 45–48
  difference, 25, 34 ff., 41, 45, 98, 100
  low temperature, 35, 38, 123–124
  photodissociation, 38, 44
  reduced *minus* oxidised, 26, 34 ff., 41, 45
  reflectance, 123–124
Split beam, 5, 25, 41, 97, 98
Stirrer, 39
Stray light, 8, 15, 30, 117, 118

Thermocouple, 39
Thioguanosine, 162–163
Thioinosine, 94, 95, 145, 148
Thymol blue, 144
Time constant, 141
Transient state, 112, 149
Transmittance, 6, 8
Trichloroacetic acid, 58, 60, 67, 83, 84, 85
Triton X-100, 59, 62, 133
Tryptophan, 93, 105, 141
Turbidity, 6, 9, 25, 50, 58, 76, 85, 93, 97, 112, 120

Ultraviolet absorbtion,
  of proteins, 58, 63
  of nucleic acids, 64

V_m, 68, 80
Valinomycin, 122, 125–130

Warm up time, 7